Measurement Techniques and Practices of Colloid and Interface Phenomena

Masahiko Abe
Editor

Measurement Techniques and Practices of Colloid and Interface Phenomena

Springer

Editor
Masahiko Abe
Research Institute for Science and Technology
Tokyo University of Science
Noda, Japan

Translation from the Japanese language edition: *GENBA DE YAKUDATSU KOROIDO KAIMEN GENSHO NO SOKUTEI NOUHAU*, © Nikkan Kogyo Shimbun 2016. Published by Nikkan Kogyo Shimbun Ltd. All Rights Reserved.

ISBN 978-981-13-5930-9 ISBN 978-981-13-5931-6 (eBook)
https://doi.org/10.1007/978-981-13-5931-6

Library of Congress Control Number: 2019933910

This Springer imprint is published by the registered company Springer Nature Singapore Pte Ltd.
The registered company address is: 152 Beach Road, #21-01/04 Gateway East, Singapore 189721, Singapore

Foreword for English Translation

Colloid and surface science is related to every phenomena of materials encountering each other, but most of the case it was overlooked or ignored in daily life because handling of materials are mostly in the bulk or mass. While recent popularization of nano-technology and expectation of its utilization is remarkable and there are many people facing to measure the properties of materials at nanometer order which is rather complicated and not easy to be accustomed.

Among the scientists who have been working for more than a hundred years in the colloid and interfacial science, people always talked about nano and developed many instrumentation in this field. Professor Abe's laboratory at Tokyo University of Science is one of the largest labs fully equipped with most of the instruments required for surface chemical measurement. To keep such activity, there are depths of expertise and resources. This is the reason Professor Abe published "Measurement Techniques and Practices of Colloid and Interface Phenomena" in 2016 and assembling training course to utilize the book for the people in the industry.

This English translation is planned to spread our experience out of Japan to help people interested in the colloid and surface science. This book is not comprehensive, rather aimed to provide tips and hints to find solutions when faced to the difficulties of measurement.

The unique value of nano is not just its size. If you take a gold cube, it is mostly bulk with comparatively little surface area. But if you dice that golden cube into many smaller pieces, the mass stays the same while the amount of surface area increases greatly. That surface area is full of energy – and full of possibilities (like purple gold particles) for the scientists working in nano.

We hope this English translation find its own way to help people working on the nanotechnology and students trying to get into the surface science.

Department of Pure and Applied Chemistry
Tokyo University of Science, Tokyo, Japan
August 2018

Kazutami Sakamoto

Preface

(This preface is translated from original Japanese publication)

This book “Measurement Techniques and Practices of Colloid and Interface Phenomena” has been prepared by scientists actively contributing for the advancement of colloid and surface science and technology who have close acquaintances with me as professor at Research Institute for Science and Technology, Tokyo University of Science. This book was planned to provide concise guidance to the young scientists entering this field on the measurement of typical physico-chemical properties relating to colloid and surface science.

This book is structured by starting introductions to the field, introduction section and followed by 19 chapters relating to measurement techniques specified to each basic properties and measurement as follows; Chapter 1: Basics of Surface Chemistry, Chapter 2: Static Surface Tension, Chapter 3: Dynamic Surface Tension, Chapter 4: Surface Pressure, Chapter 5: Surface Viscosity, Chapter 6: Interfacial Tension between Water and Oils, Chapter 7: Quartz Crystal Microbalance with Dissipation monitoring (QCM-D), Chapter 8: Atomic Force Microscope (AFM), Chapter 9: Static Light Scattering (SLS), Chapter 10: Dynamic Light Scattering (DLS), Chapter 11: Solubilization by Micelle, Chapter 12: Rheology, Chapter 13: Freeze-Fracture Transmission Electron Microscopy, Chapter 14: Cryo-Transmission Electron Microscopy, Chapter 15: Zeta(ζ)-potential for Micelle and Microemulsion, Chapter 16: Electron Microscopy Observation of Solid Particles, Chapter 17: Gas Adsorption on Surface of Solid Materials, Chapter 18: Contact Angle Measurement for Solid Surface, Chapter 19: Quality and treatment of water for experiment.

Each chapter consists of sub sections including *Introduction, What You Get, Essentials and Tips, Understanding Your Data, What to Look Out For, Useful Hints and additional sub sections for some chapters* from a beginner’s view point.

Chapter 17 explains how to prepare physico-chemically pure water, especially for nanometer level measurement, to avoid biological contamination.

I hope this book fits the reader's interest and helps to support their research advancement.

Noda, Japan
April 2016

Masahiko Abe

Contents

Chapter 1
Basics of Surface Chemistry

Masahiko Abe

Abstract This chapter deals with the basics of surface chemistry such as the difference between interface and surface, the relationship between interface and colloid, various types of colloid dispersions with gas-liquid interface, gas-solid interface, liquid-liquid interface, liquid-solid interface, and solid-solid interface, reactions at interfaces, and the difference between surfaces and internal areas. Fundamentals of surfactants (surface-active agents) based on their classification, solubility, Krafft point, cloud point, salting out, and salting in are discussed in brief.

Keywords Interface · Surface · Colloidal dispersion · Surfactant · Krafft point · Cloud point

1.1 Introduction

Surface chemistry and colloid chemistry are closely related fields that study colloid dimensions of 10^{-9}–10^{-5} m, a scope shared with the highly focused field of nanotechnology. When studying a two-phase system, the border between the two phases is called an interface, and the interface area between the phases can be maximized by decreasing the phase of one substance and dispersing it in the other continuous phase. Surfactants are commonly used to create such dispersed systems. In this book, we will investigate several important analytical methods among the various methods used in research dealing with surfactants.

M. Abe (✉)
Research Institute for Science and Technology, Tokyo University of Science, Noda, Japan
e-mail: abemasa@rs.noda.tus.ac.jp

M. Abe (ed.), *Measurement Techniques and Practices of Colloid and Interface Phenomena*, https://doi.org/10.1007/978-981-13-5931-6_1

1.2 Definition of Interface

An interface is defined as the boundary between two phases with different properties, and if either of the phases is a vacuum or a gas, this interface is also called a surface. A common example is a glass of water, where the water contacts the air and thus the interface is called a surface. The boundary between two phases is actually thought to be a continuously changing state where the two phases both penetrate into each other. This means that even though the interfacial layer is only a few nanometers thick, it is not a simple two-dimensional interface but a three-dimensional interfacial zone. In thermodynamics, an interface is treated as a single phase. As such, the free energy of the interface must be positive for the interface to exist between two phases. If the free energy is equal to or below zero, the interfacial area continuously expands and eventually disappears.

Surfaces and/or interfaces exist in every material, and the properties vary greatly between the interface and within each phase. The surface of pure water has a larger amount of free energy compared to inside the phase (bulk), so the surface will try to make its area minimal, and the concentration of solutes at the surface of an aqueous solution is different from that of the bulk of the solution since the solutes are adsorbed on the surface. In this case, it is noteworthy that when the ratio of the interface area is extremely large against the volume of the system, the properties of the interface can define the properties of the whole substance system. In fact, the total area of interfaces is larger than what we tend to imagine. For example, if one side of a regular hexahedron, i.e., a six-faced cube made of squares, is 1 cm, the surface area is $1 \times 6 = 6$ (cm^2). However, if 1000 regular hexahedrons are inside the same volume with the sides one tenth of the original hexahedron, the total surface area is 60 cm^2. If the sides are shortened into much smaller hexahedrons with 1 nm sides, the total surface area becomes 6×10^7 cm^2, equivalent to 6940 m^2, which is larger than 1.4 acres. As such, when the total surface area is increased, the two same materials with the same substances can have significantly different properties. For example, a steel nail of visible size is not flammable, but fine particles of steel can be oxidized rapidly and thus is flammable. The precious metal we know as gold is "golden," but as fine particles it can have a reddish hue, and sometimes it is not a stable metal.

Substances generally exist in either a solid phase, liquid phase, or gas phase, and interfaces can be made from these phases as follows:

1. Gas-liquid interface
1. Liquid-liquid interface
1. Gas-solid interface
2. Liquid-solid interface
3. Solid-solid interface

Note that gases mix together and do not create a gas-gas interface. Additionally, liquid crystals and other recently found states can be considered a phase and can be added to the combinations of interfaces.

1.3 Interface and Colloid

When a phase in contact at the interface with another phase is minimized to the range of 1 nm–1 μm, the system is called a colloidal dispersion. The fine particles in this colloidal dispersion is called the dispersed phase or dispersoid, and the surrounding phase (continuous phase) is called the dispersing medium.

A colloidal dispersion is a system where a dispersed phase (dispersed particles) that is larger than a molecular dispersion and smaller than a coarse dispersion is in the dispersed medium, and the particles can be spherical, rod like, planar, polyhedral, or coiled. For example, colloids such as smoke have solid particles in a gas phase, fog has liquid particles in a gas phase, milk has liquid particles in a liquid phase, mud has solid particles in a liquid phase, sponges have gas particles in a solid phase, gelatin has liquid particles in a solid phase, and colored glass has solid particles in a solid phase. Table 1.1 shows various types of colloid dispersions.

The examples mentioned above are categorized by the type of dispersed phase and dispersed medium, but colloids can also be categorized by types and states of the dispersed particles, namely molecular colloids, association colloids, and dispersion colloids. Molecular colloids are systems such as gelatin or proteins in aqueous phases or polystyrene in benzene solutions, where the dispersed particles are polymers or polymer ions dispersed in liquids with low molecular weight. Polymers naturally dissolved in solvents were previously called eucolloids, but this term is rarely used today. Association colloids are systems where molecules such as surfactants or dyes are dispersed in a solution as molecular associations with tens or hundreds of molecules, and the molecular associations' dissociation and association are reversible when diluted, concentrated, heated, or cooled. Molecular colloids and association colloids are thermodynamically equilibrium systems that are reversible and thus are called reversible colloids or stable colloids. Dispersion colloids are systems where substances that do not normally disperse in the medium are dispersed by special treatment and are called irreversible colloids or unstable colloids. These irreversible colloids rely on interfacial tension and/or interfacial electrical charge, so molecular colloids or association colloids are often added to as protecting colloids to stabilize the system. Colloids can also be categorized by the affinity of the dispersed phase and dispersing medium, and in this categorization, association colloids are

Table 1.1 Various types of colloid dispersions

Dispersion media	Dispersoid	Dispersion system
Gas	Liquid	Fog, cloud, aerosol, gas colloid
	Solid	Vapor
Liquid	Gas	Bubble
	Liquid	Emulsion (milk, butter)
	Solid	Suspension (mud water, paint)
Solid	Gas	Floatstone, sponge, sea sponge, solid colloid
	Liquid	Silica gel having water
	Solid	Composition metal

called lyophilic colloids. Lyophilic colloids are called hydrophilic colloids or lipophobic colloids if the dispersed medium is water. Colloid solutions show many characteristic kinetic and optical properties, but one of the most prominent is the Tyndall phenomenon.

Chemists, focusing on colloid research, study colloids as three-dimensional systems as particles mentioned above but also study two-dimensional colloid dimensions as fibers and filaments and single-dimensional colloid dimensions as membranes.

As Table 1.1 shows, the interfaces in 1–5 are activated, and this phenomenon is called surface activation or interface activation. Previously, notable change in surface tension or interfacial tension due to small amounts of solutes was called interfacial activation, but there are phenomena where dispersion is caused without such changes. In other words, the change in the various properties of interfaces or surfaces is interfacial activation, and naturally this can be positive interfacial activation and negative interfacial activation. Foaming and defoaming, dispersion and coagulation (or flocculation), emulsification and demulsification, and hydrophobication or lipophobication on a solid surface are examples of such activation. Substances that show interfacial activation are called surface-active agents (surfactants), detergents, or amphiphilic compounds. The term for surfactants can change depending on the functions of its application. For example, surfactants are called emulsifying agents, dispersing agents, detergents or cleansers or washing agents, and solubilizing agents.

1.4 Reaction at Interfaces

Interfaces can be studied as an area of chemical reactions. Chemical reactions are generally categorized into homogeneous reactions and heterogeneous reactions, where homogeneous reactions are reactions that occur in single phases, such as reactions between gases and liquids in an equally mixed or solvent system, or decomposition (i.e., degradation, cracking breakdown, resolution reactions) in a single phase. Other reactions are all heterogeneous reactions, or reactions that occur in two or more separate phases, at the interface and are called interfacial reactions. Reactions that are vital in industrial application, such as electrode reaction, catalytic reaction, adsorption, and crystal growth, are all interfacial reactions. Additionally, interfacial reactions can be categorized by the type of phases as shown below:

1. Gas-liquid interface reactions: reactions such as same-molecule gas-liquid phase transition, synthesis reaction at gas-liquid interfaces
2. Gas-solid interface reactions: reactions such as reactions on solid surfaces as catalysts
3. Liquid-liquid interface reactions: reactions such as emulsion polymerization reaction, reactions at biomembranes

4. Liquid-solid interface reactions: reactions such as electrode reactions, adsorption reactions, and crystallization reactions
5. Solid-solid interface reactions: reactions such as formation of solid solutions

To analyze the reaction mechanism in such elements, it is vital to focus and select the interface (and omit information irreverent to the interface) and analyze the system in real-time. There are some restrictions when analyzing the static properties of the interface in this manner. X-rays and electron beams are vital in analyzing solid surfaces, but these methods are ex situ (static) analytical methods and cannot trace the dynamic changes of interfacial reactions on an interface. Additionally, these methods usually cannot be used for liquids. Real-time methods for monitoring interfacial reactions that can be used in situ and do not rely on the atmosphere such as light probing methods and scanning probe microscopes are gaining attention.

1.5 Surfaces and Internal Areas

The atomic state is not always the same on the surface and in the bulk of a single solid. For example, when a solid in air is cut parallel to the inner atomic plane, a new gas-solid interface is formed. The gas-solid interface consisting of the solid atoms on the new surface is unstable and hence it increases the surface energy. To stabilize the surface (and decrease the surface energy), the atomic arrangement must be realigned by changing the vertical interlayer spacing, resulting in a different alignment compared to the inner atoms. Additionally, surface molecule realignment can also occur with surface reactions of gases and liquids adsorbed existing solid surface. Properties sensitive to the structural change of solids such as atomic vibration, chemical properties, optical properties, electronic properties, and/or magnetic properties change depending on the state of the surface, thus making the surface a vital area that affects the properties of a substance. Furthermore, as mentioned above, as the size of the solids become smaller, the property of the surface strongly influences the property of the solids and can even activate them. The surface of a solid is also affected by temperature and/or atmosphere, and in some cases small amounts of impurities inside the solid concentrate on the surface.

1.6 Fundamentals of Surfactants

1.6.1 Classification of Surfactants

Surfactants are substances that significantly change the properties of surfaces and interfaces when small amounts are added to the system. This significant change in the surface or interface is caused due to the dual property of surface activation, one being a positive activation by increasing the interfacial area by increasing the affinity of the two phases of the interface and the other a negative activation when the affinity is decreased to decrease the area of the interface. The following are some examples of each activation:

Positive interfacial activation: Dispersion, penetration, wetting, bubbling, solubilization, emulsification, etc.
Negative interfacial activation: Aggregation, repelling, defoaming, demulsification, etc.

There are many naturally derived and artificial substances that are surface active, and these surface-activating agents used in both personal/household and industrial/institutional are called surfactants, and naturally derived surfactants are called biosurfactants.

Surfactants are classified into types below depending on their hydrophilic group:

1. Anionic type
2. Cationic type
3. Amphoteric or zwitterionic type
4. Nonionic type

Additionally, surfactants are classified into the types below depending on their hydrophobic group:

Hydrocarbon type
Fluorocarbon type
Organosilicon type
Saccharide or glucoside type
Others

Surfactants have a structural characteristic with at least two kinds of functional groups in either the hydrophilic group (or lipophobic group) or the lipophilic group (or hydrophobic group), and common surfactants are categorized into the six types below depending on the number of amphiphilic groups.

1. Single hydrophilic group single lipophilic group type
2. Single hydrophilic group dual lipophilic group type
3. Dual hydrophilic group single lipophilic group type
4. Dual hydrophilic group dual lipophilic group type
5. Multi hydrophilic group multi lipophilic group type
6. Single hydrophilic group heterogenous hydrophobic group type

Note: Gemini surfactants attracted global attention since the 1990s [1, 2] *(usually categorized in type 4 but sometimes as type 5), and although the molecular weight of these surfactants is about twice of type 1 (normal monomeric surfactants), the critical micelle concentration (cmc) is extremely small with 1/100–1/1000 compared to monomeric surfactant, and also these surfactants have a low Krafft point and show a strong water surface tension decreasing potential.*

Unlike other chemical substances, interface activating substances (surfactants) show a critical change in solubility when the concentration of the solution is

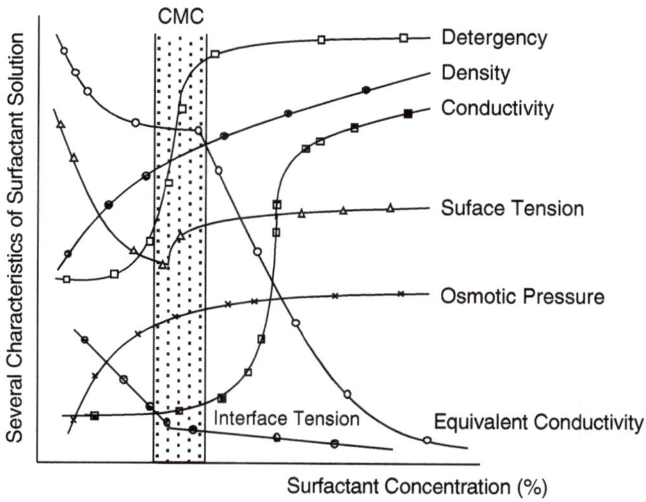

Fig. 1.1 Various physical characteristics of surfactant with its concentration change

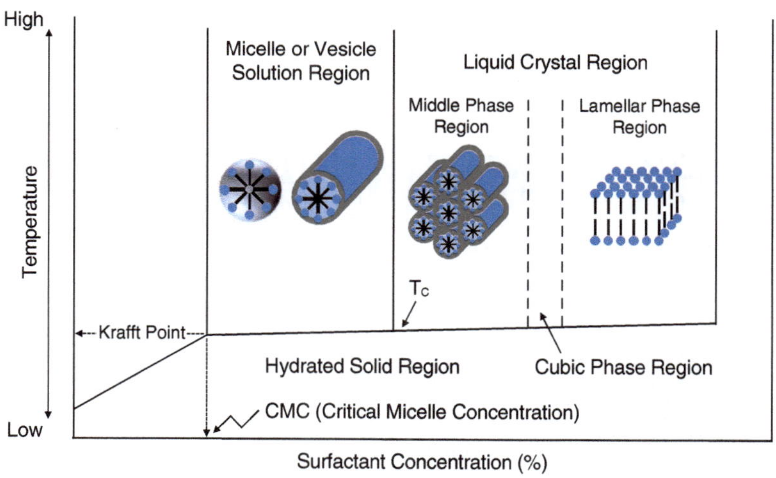

Fig. 1.2 Various colloidal assemblies formed in ionic surfactant/water systems

increased under a fixed temperature, and at this concentration (cmc), the interfacial activation also shows a significant change (Fig. 1.1). Especially as shown in Fig. 1.2, if the temperature of the system is higher than the Krafft point, at certain concentrations (cmc), the system shows colloidal assemblies such as micelles, vesicles, and liquid crystals.

Figure 1.3 shows colloidal assemblies made of surfactant molecules. The shapes of these associations show a correlation to the chemical structure of the surfactant to a certain extent and can be calculated using the critical packing parameter, or CPP, using the cross-sectional area per molecule of the surfactant's hydrophilic group on

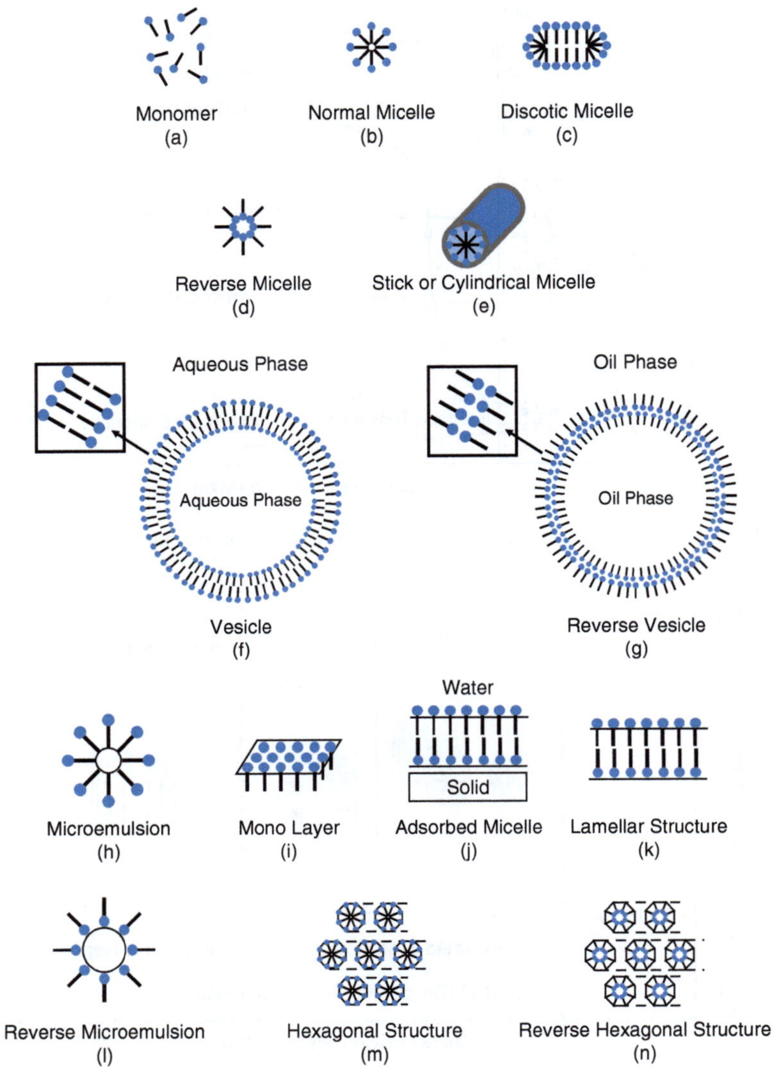

Fig. 1.3 Various colloidal assemblies of surfactants

surface or interface (a_0), the volume of the hydrophobic group in the surfactant molecule (v), and the length of the same hydrophobic group (l_c) as CPP = $v/(a_0 \times l_c)$. The correlation of these parameters is shown in Fig. 1.4. It is understood that spherical micelles are formed when the value is small (lower than 1/3), vesicles or planar bimolecular layers are formed when the value is approximately 1, and if greater than 1, reverse micelles are formed.

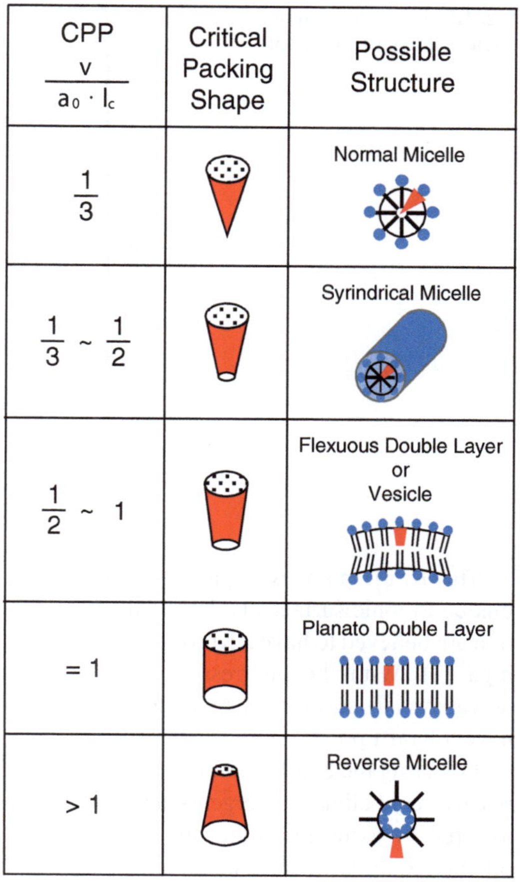

Fig. 1.4 Relationship between critical packing parameter and possible structure of colloidal assemblies

1.6.2 Solubility of Surfactants

1.6.2.1 Krafft Point (KP)

The aqueous solubility of surfactants differs greatly from that of other chemical substances. As shown in Fig. 1.5, surfactants form a colloidal dimensional molecular cluster called micelles (II) when it reaches a certain temperature, and the solubility increases greatly as these clusters dissolve into water while maintaining its structure. This temperature is called the Krafft point or Krafft temperature.

The Krafft point is a physical quantity unique to each surfactant and is regarded the melting point of the surfactant's hydrated solid. Below the Krafft point, the surfactant dissolves as a monomer, and the monomer concentration of the micelle solution does not change with the concentration of the surfactant.

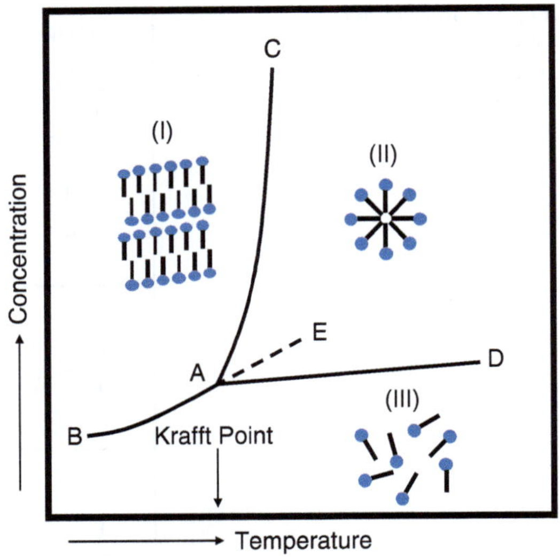

Fig. 1.5 Phase diagram of dilute solution of surfactant

The Krafft point was previously considered an indicator for water solubility unique to ionic surfactants, but nonionic surfactants that do not show ion dissociation are believed to have a Krafft point in temperature ranges over 0 °C if the purity is high [3]. With the progress in surfactant synthesis and refinement techniques, nowadays all surfactants regardless of their type should be treated to generally have a Krafft point.

Lowering the Krafft point is an extremely important factor in practical application (such as cold climate detergents and North Sea oil recovery). The Krafft point can be lowered by focusing on the surfactant structure and change factors such as hydrophobic groups with unsaturated bonds or branched chains, introducing polyethylene oxide bonds, changing counterions, and mixing other surfactants. Addition of alcohols lowers the melting point and lowers the Krafft point.

1.6.2.2 Cloud Point

Typically, nonionic surfactants are dissolved in water as micelles when temperature is higher than the KP due to hydration of its hydrophilic group (polyoxyethylene groups or polyoxypropylene groups), but when it reaches a certain temperature, the hydrophilic group is dehydrated and becomes less soluble, and at this point the surfactant molecules separate with the phases, and the solution becomes clouded. The cloud point is the lowest temperature where such liquid-liquid phase separation occurs, and unlike the Krafft point, the cloud point changes depending on the entire concentration of the surfactant in the solution. The cloud point usually is higher when the number of hydrophilic groups is larger in the surfactant molecules and decreases when inorganic electrolytes (salts) are added. The effect of these salts is deeply connected to the Hofmeister effect of the salts' lyotropic series, where the order is $SO_4^{2-} > Cl^- > Br^- > NO^{3-}$. Thus, the cloud point decreases more if the *salting out* effect is stronger. Like KSCK and urea, substances with *salting in* effect can increase the cloud point.

The cloud point is an indicator of the affinity of nonionic surfactants against water and is also deeply related to the HLB, or hydrophilic lipophilic balance. In recent years, cloud points have been observed in ionic surfactants with polyoxyethylene groups [4] or mixed systems of anionic surfactants without polyoxyethylene groups and cationic surfactants [5].

As seen from these observations, it is important to understand that all surfactants, regardless of its ionicity, have a Krafft point in lower temperature zones and a cloud point in higher temperature zones.

References

1. F.M. Menger, C.A. Liitau, J. Am. Chem. Soc. **113**, 1451 (1991)
2. R. Zana, Adv. Colloid Interf. Sci. **97**, 205 (2002)
3. D.J. Mitchell, G.J.T. Tiddy, L. Waring, T. Bostock, P. Mcdonald, J. Chem., Faraday Trans. I **79**, 975 (1983)
4. M. Abe, D. Schechter, R.S. Schechter, W.H. Wade, U. Weerasooriya, S. Yiv, J. Colloid Interface Sci. **114**, 342 (1986)
5. Y. Nakama, F. Harusawa, I. Murotani, J. Am. Oil Chem. Soc. **67**, 717 (1990)

Chapter 2
Static Surface Tension

Masahiko Abe

Abstract Surfactants (surface-active agent) are defined as substances capable of changing the surface or interfacial properties significantly. Addition of small amounts of surfactants into water decreases surface tension significantly. Namely, it is not called a surfactant if it does not decrease the surface tension of aqueous solutions. This chapter introduces the physical definition of surface tension and methods to measure it, such as Wilhelmy plate method and Du Nöuy method. The most popular methods and essential points for the measurement are explained. Surface tension measurements enable to determine the critical micelle concentration (cmc) of the surfactant, calculate the amount of surfactant adsorbed on the surface, and to determine the miscibility of two surfactants in mixed micellar solutions.

Keywords Surface excess energy · Surface tension · Critical micelle concentration · Surface adsorption amount · Miscibility

2.1 Introduction

Liquid surface in a large container gives a flat surface, though intrinsic liquid surface does not always behave so, as the shape of dew drops on leaves or on solid surface shows. This is because surface molecules of a pure liquid has larger free energy than bulk molecules, so that the surface shape tends to be spherical to minimize surface excess energy with the least number of molecules on the surface. Surface tension is a force to shrink the surface with dimension of force per unit length (mN/m), which is equal to the surface free energy per area (mJ/m^2). Figure 2.1 shows the physicochemical means of surface tension. Let's focus on the single molecule in the bulk (shown by a black dot in Fig. 2.1a). This molecule is surrounded by the same molecules around with same intermolecular force from all directions. As a result, the total amount of intermolecular force with surrounding molecules become zero. When these molecules

M. Abe (✉)
Research Institute for Science and Technology, Tokyo University of Science, Noda, Japan
e-mail: abemasa@rs.noda.tus.ac.jp

M. Abe (ed.), *Measurement Techniques and Practices of Colloid and Interface Phenomena*, https://doi.org/10.1007/978-981-13-5931-6_2

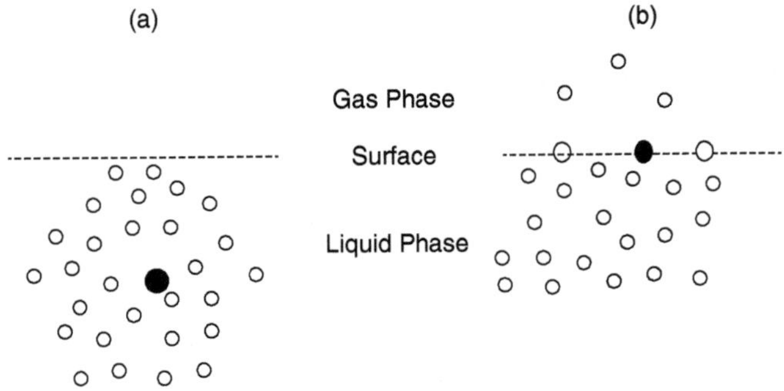

Fig. 2.1 Differences of molecular density at gas/liquid interface. The black dot shows single molecule in the bulk (**a**) or at the surface (**b**)

are brought closer from infinity, potential energy will be reduced to negative corresponding to the total amount of intermolecular forces with surrounding molecules. Thus each single molecule in the bulk (shown by black dot in Fig. 2.1a) is stable.

On the contrary, a molecule at the surface shown as a black dot in Fig. 2.1b has smaller intermolecular forces with molecules in the gas phase toward the outside of surface. As a result, the total intermolecular forces can not be cancelled to zero so that surface molecule has excess higher energy than the molecule in the bulk. This reflects to the required work force to make new surface by bringing bulk molecule to the surface. This work force, corresponding to the surface excess energy and, namely, surface tension, is smaller than the difference of potential energy between bulk molecule and surface molecule, because of the larger entropy of surface molecule. As a result, surface tension (surface energy) is larger for the molecule with strong intermolecular interaction. Surface tension of solid material, which is called as critical surface tension, comes from the same origin, and it can be measured by the contact angles of liquids dropped on the solid surface as explained in Chap. 18. Critical surface tension can be empirically calculated from the sum of contact angle dependency on the length of hydrophobic groups for liquid A with only dispersion force component, liquid B with both dispersion force component and polar component, and liquid C with dispersion force component, polar component, and hydrogen bond component.

As explained above, surface tension (γ) is expressed with dimension of force per unit length (mN/m) and can be also converted as

$$
\begin{aligned}
(\gamma)\ [\mathrm{mN/m}] &= [\mathrm{force}]/[\mathrm{length}] = [\mathrm{force}] \times [\mathrm{length}]/[\mathrm{length}] \times [\mathrm{length}] \\
&= [\mathrm{work}]/[\mathrm{area}] \quad \text{or} \quad [\mathrm{energy}]/[\mathrm{area}]
\end{aligned}
$$

Thus, γ(mN/m) can be expressed by (mJ/m^2) which corresponds to the work to make a new unit area. As surface tension reflects the strength of intermolecular interaction, it is closely related to the chemical structure as shown in Table 2.1. Surface tension is dependent on the temperature as temperature increase corresponds

Table 2.1 Surface tension for liquid materials

Material	Surface tension (mN/m)
Water	72.75
Glycerin	63.4
Benzene	28.9
Toluene	28.5
Acetone	23.7
Ethanol	22.8
Methanol	22.6
n-Hexane	18.4
Ethyl ether	17.0

to the incremental thermal motion and requires less work to bring bulk molecule to the surface. Also, surface tension depends on the solute concentration at the surface.

There are various methods for static surface tension measurement. The most popular methods are drop weight method (stalagmometric method), capillary rise method, maximum bubble pressure method, Du Nöuy method, Wilhelmy plate method, and pendant drop method. Those methods measure static surface tension at the equilibrium which closely relates to the forming capacity. The foam stability by drainage velocity depends on the dynamic surface tension, which can be measured by oscillation jet method and Hiss method (Chap. 3). Foam stability also depends on the surface viscosity (Chap. 5).

When new molecule as a surfactant is synthesized, it is a common practice to measure the level of surface tension reduction for water to evaluate the potential of this molecule. It can be found in the traditional textbook that says "addition of tiny amount of surfactant (surface active agent) can reduce water surface tension significantly," as a reflection common perception has been that not to say surfactant without surface tension reduction. As time passes by, polymer surfactant with poor surface tension reduction but superior dispersibility of materials in the water was developed, which made surfactant definition been changed as "material capable of changing surface or interfacial properties significantly." As explained later, critical micelle concentration (cmc) can be determined from the plot of surface tension vs concentration. In order to get accurate cmc, the relationship between equivalent conductivity and concentration is desirable, though it would not be applicable for nonionic surfactant by currently available electrodes. The methods explained here for the surface tension measurement are the most convenient way because they are independent from the ionic properties of surfactant.

2.2 What You Get

1. The cmc in water can be determined regardless of the types of surfactant.
2. Surface adsorption amount of surfactant in the solution (mostly in water) can be measured.

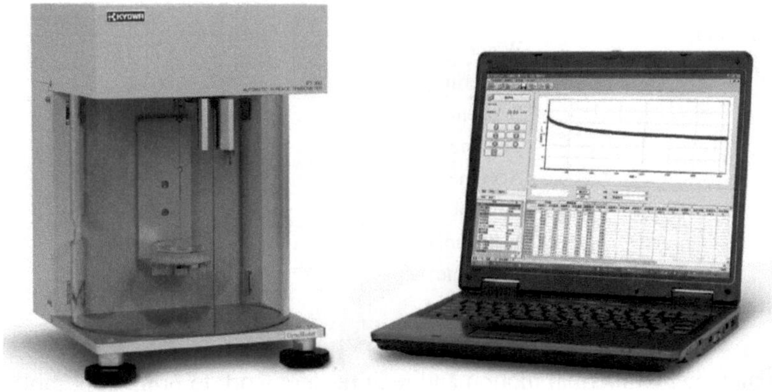

Fig. 2.2 Surface tensiometer DY-300. (Photo credit Kyowa Interface Science Co. Ltd.)

3. Molecular cross-sectional area at the air/liquid interface can be calculated.
4. Miscibility conditions of binary surfactant in aqueous solution can be determined.

2.3 Method

The most popular methods are the Wilhelmy plate method and Du Nöuy method. The pendant drop method, drop weight method (stalagmometric method), capillary rise method, and maximum bubble pressure method are commonly used. To get detailed methodology, operation manual for each method should be referred as each has their own principle. For the Wilhelmy plate method, one of the most popular methods, it is very important to prepare absolutely clean platinum or glass plate by complete cleansing and drying and avoiding contamination and surface disruption of sample solution by breath (Fig. 2.2). If fluorinated surfactants are the object of measurement, wetting control of the plate surface by chemical treatment is necessary.

2.4 Essentials and Tips

Du Nöuy method with a platinum ring has been the most widely used, even out of chemical industry, because of its simple and easy handling. The Wilhelmy plate method with a platinum or glass plate has been used more like professional field for its accuracy. The drop weight method (stalagmometric method) is also popular, and will be explained in the next chapter for the dynamic surface tension measurement.

Table 2.1 shows the surface tension of typical liquids. Data acquired by the method explained are surface tension vs surfactant concentration as shown in Fig. 2.3. In order for scientific analysis, concentration should be expressed by

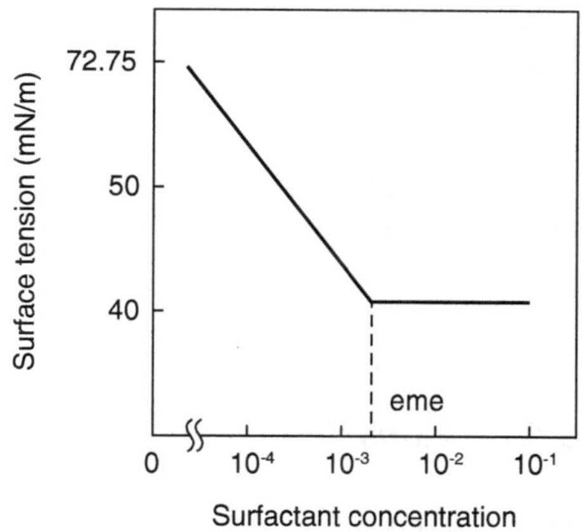

Fig. 2.3 Relationship between surface tension and surfactant concentration in water

molar concentration (mol/l). Critical micelle concentration (cmc) can be determined from the cross section as shown in Fig. 2.3 over which surface tension becomes constant, which can be explained as a saturation concentration of monomeric surfactant solution. Smaller in cmc means higher in surface activity. For the single-tailed surfactant, cmc is 10^{-4}–10^{-3} mol/l for ionic, regardless of ionic type, and 10^{-5}–10^{-6} mol/l for nonionic surfactants. Gemini-type surfactants with double tail cmc sometime decreases to 2–3 orders smaller. Ionic surfactants with EO (polyoxyethylene) tend to show lower cmc as nonionics.

2.5 Understanding Your Data

As clear from Eq. 2.1, the adsorption amount of the surfactant at the air/water interface can be calculated from the surface tension slope toward cmc. The important point here is to use the slope as close to the cmc. In case the slope of surface tension vs concentration is not in uniformity but stepwise, the range of concentration can be divided to discuss adsorption and desorption of surfactant.

$$\Gamma = \frac{1}{A} = -\frac{1}{R}\frac{\mathrm{dr}}{d\ \ln C} \tag{2.1}$$

where Γ is the amount of surfactant adsorbed on the surface per unit area, R is gas constant, γ is surface tension, C is molar concentration (mol/l), and A is surface area per single surfactant molecule adsorbed.

As this equation stands for at the equilibrium, the surface tension value should be an equilibrium value. As surface tension is dependent on time duration and sometime requires quite a long period to reach equilibrium, special care should be paid for the measurement. In case concentration is not low enough to use mol/l, activity should be used instead.

It is realized that mixed surfactant system performs better than single and became quite popular for practical use. Two books are published for these characteristics of mixed surfactant system [1, 2]. Theoretical equations for the interaction between surfactants in aqueous mixed surfactant solution have been proposed by Funasaki [3], Rubingh [4], and Motomura [5]. A summary of representative two examples is explained here based on the Funasaki's equation for the binary mixed surfactant micelles [6].

2.5.1 Example 1

Mixed surfactant solutions[6] at 5.0×10^{-3} M consist either of sodium 3,6, 9-trioxaicosanoate (ECL from Nikko Chemical) or sodium dodecyl sulfate (SDS) as anionic surfactant with hexadecyl polyoxyethylene ether ($C_{16}H_{33}O(EO)_{10}H$; POE from Nikko Chemical) as nonionic surfactant. Composition of ECL (X_{2m}) and POE (X_{1m}) in the mixed micelle over cmc is calculated in Eqs. 2.2 and 2.3.

$$X_{2m} = \frac{C_1 * X_2 - C_{12} * X_{2b}}{C_t - C_{12}} \tag{2.2}$$

$$X_{1m} = 1 - X_{2m} \tag{2.3}$$

where C_t is the total surfactant concentration in the mixture (mol/l; M), C_{12} is the concentration of the total surfactant in the bulk phase (mol/l; M), X_2 is molar ratio of ECL in the mixed solution, and X_{2b} is molar ratio of ECL in the bulk phase.

Data required for the calculation can be retrieved from Fig. 2.4 as surface tension vs concentration relationship for each single component and Fig. 2.5 as surface tension vs concentration relationship for the binary mixed solutions at different ratio and Fig. 2.6 as surface tension (—○—) or concentration (—●—) vs mole fraction of ECL in the bulk phase obtained from Figs. 2.4 and 2.5. C_t in Eq. 2.2 is concentration

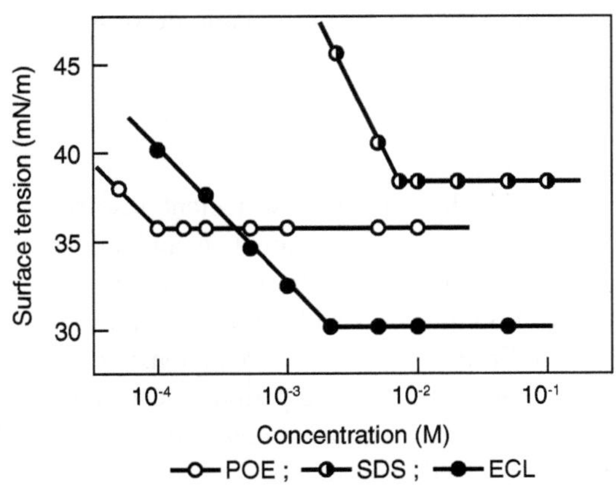

Fig. 2.4 Relationship between surface tension and concentration for single surfactant

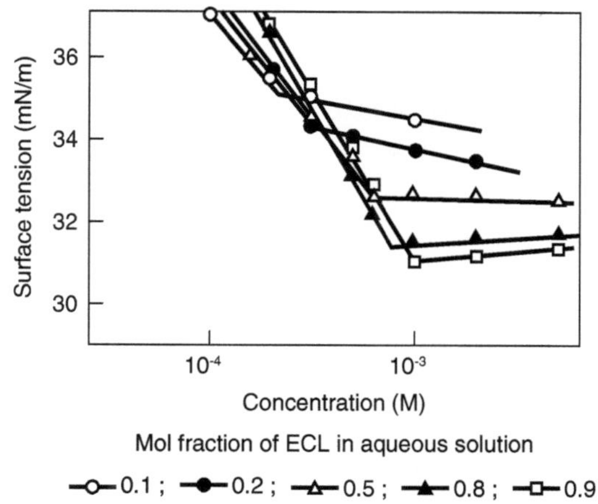

Fig. 2.5 Relationship between surface tension and concentration for binary surfactant mixture (ECL-POE) (cross point reflects cmc)

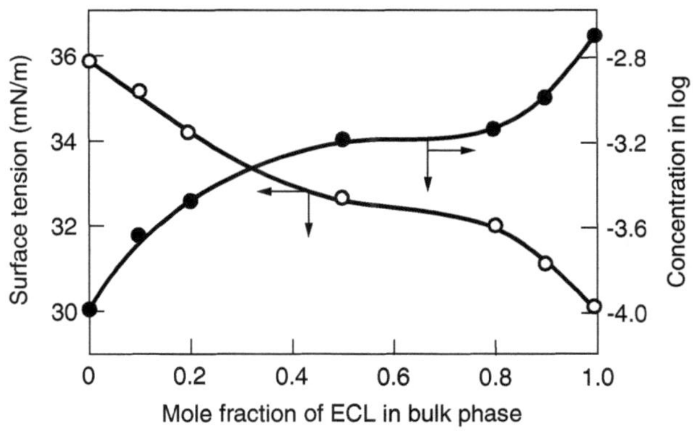

Fig. 2.6 Relationship between surface tension (—○—) or concentration (—●—) and mole fraction of ECL in the bulk phase for ECL-POE system

of mixed solution (mol/l), X_2 is composition of mixed solution, and X_{2b} and C_{12} can be determined from this figure. Applying these data to Eq. 2.3, the molar ratio of ECL or SLS in the micelle can be calculated.

The other example has been reported by Motomura et al. [5], where molar fraction of binary mixture can be calculated by Eq. 2.4 [7]. We have adapted this to the binary mixture of N^{α}, N^{α}-dimethyl-lauroyl lysine (DMLL) as amphoteric surfactant and alkyl-polyoxyethylene-ether ($C_nPOE_{20;}$ $n = 12, 14, 16, 18$) as non-ionic surfactant.

$$X_{\mathrm{DMLL}}^{M} = X_{\mathrm{DMLL}} - \left(\frac{X_{\mathrm{C}_m\mathrm{POE}_{20}} {}^{*} X_{\mathrm{DMLL}}}{\mathrm{cmc}}\right) * \left(\frac{\delta_{\mathrm{cmc}}}{\delta X_{\mathrm{DMLL}}}\right)_{T*P} \tag{2.4}$$

where X_{DMLL} is molar fraction of DMLL, X_{CnPOE020} is molar faction of C_nPOE_{20} in the total solution, and cmc corresponds to the cmc of each combination.

Fig. 2.7 Relationship between cmc and composition for DMLL-C_nPOE_{20} system at 40 °C

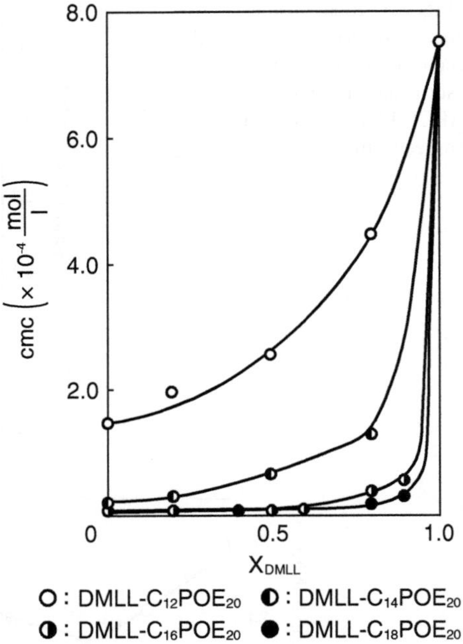

Fig. 2.8 Relationship between cmc and mol fraction of DMLL in micelle (X^M_{DMLL}) or bulk (X_{DMLL}) for DMLL-$C_{12}POE_{20}$ system at 40 °C

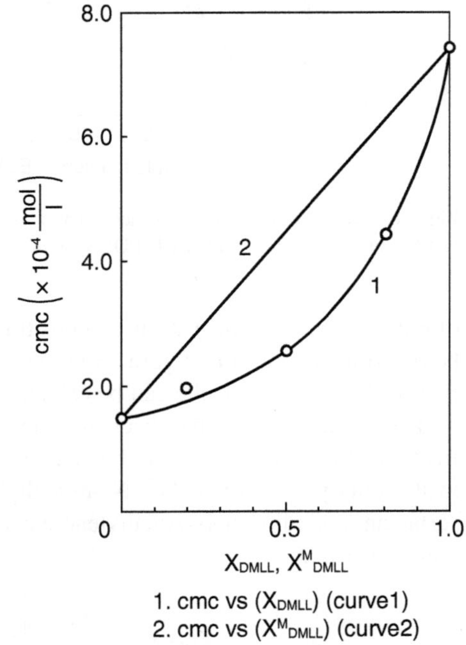

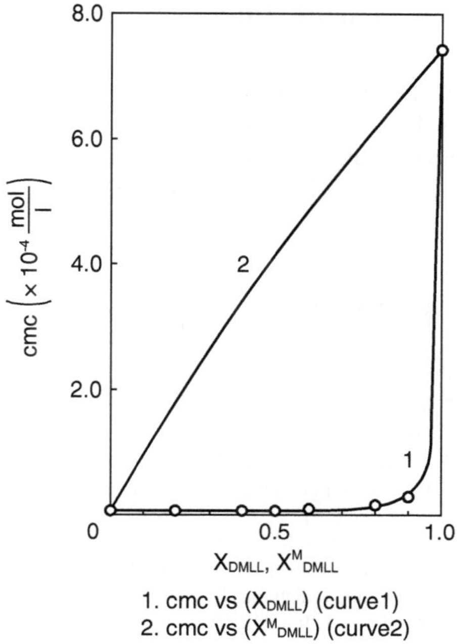

Fig. 2.9 Relationship between cmc and mol fraction of DMLL in micelle (X^M_{DMLL}) or bulk (X_{DMLL}) for DMLL-$C_{18}POE_{20}$ system at 40 °C

Figure 2.7 shows the relationship between cmc and composition (X_{DMLL}) for DMLL- C_nPOE_{20} system [7]. For the mixture of DMLL and $C_{12}POE_{20}$ (Fig. 2.8) or $C_{18}POE_{20}$ (Fig. 2.9) was plotted as cmc vs X^M_{DMLL} (curve 2) where X^M_{DMLL} was calculated from the value obtained from Fig. 2.4 together with the data for cmc vs X_{DMLL} from Fig. 2.7 as curve 1. Curve 1 shows cmc vs molar ratio of DMLL in the solution, and curve 2 shows cmc vs molar ratio of DMLL in the micelle obtained from Eq. 2.4. Motomura explained that ideality is smaller when separation of curve 1 and 2 is larger based on the phase separation model [5], and it is considered that interaction between surfactants is larger when difference from the ideality is larger [8]. Namely, larger in discrepancy between curve 1 and curve 2, interaction between surfactants is stronger.

2.6 What to Look Out for

Surface tension for pure water at a given condition is intrinsic physical constant, 72.75 mN/m at 25 °C, 1 atm. By Du Nöuy method, the value obtained is sometime not consistent, so that apparatus should be set as 72.75 mN/m to be maximum with pure water. Water should be clean from physical aspects which is prepared from deionized water after distillation and the surface swiped with silk thread filled in the large dish.

2.7 Useful Hints

Surface tension is closely related to the foaming property. Smaller in surface tension makes foaming easier as foaming is a process that makes new air/liquid surfaces. Foam stability relates to the surface viscosity which is a dynamic process, and dynamic surface tension should be used for the foam stability.

References

1. K. Ogino, M. Abe (eds.), *Mixed Surfactant Systems* (Marcel Dekker, New York, 1992)
2. M. Abe, J. F. Scamehorn (eds.), *Mixed Surfactant Systems*, 2nd edn. (Marcel Dekker, New York, 2005)
3. N. Funasaki, S. Hada, J. Phys. Chem. **83**, 2471 (1979)
4. D.N. Rubingh, in *Solution Chemistry of Surfactants*, ed. by K. L. Mittal, (Plenum Press, New York, 1979), p. 337
5. K. Motomura, M. Yamanaka, M. Aratono, Colloid Polym. Sci. **262**, 948 (1984)
6. K. Ogino, M. Abe, N. Tsubaki, J. Jpn. Oil Chem. Soc. **31**, 953 (1982)
7. K. Ogino, T. Kubota, K. Kato, M. Abe, J. Jpn. Oil Chem. Soc. **36**, 432 (1987)
8. Y. Moroi, J. Jpn. Oil Chem. Soc. **20**, 596 (1980)

Chapter 3
Dynamic Surface Tension

Taku Ogura

Abstract Like how static surface tension indicates various basic properties of surfactants, dynamic surface tension is an indicator for various features of surfactants under transient and non-equilibrium conditions such as concentration dependency or detergent foaming. Such features are key factors in industrial application. For example, determining the dynamic surface tension can be applied to understanding emulsification or wetting properties. When surfactants adsorb to interfaces, the interfacial tension reduces and correlates to these properties. In this chapter, maximum bubble pressure measurement is introduced as a standard method to determine the dynamic surface tension along with calculation methods to determine various properties of the sample surfactant from the results.

Keywords Maximum bubble pressure · Dynamic surface tension · Lifetime of bubbles · Wettability · Surface tension reduction · Surfactant

3.1 Introduction

Static surface tension measurement is indispensable to understand the intrinsic basic properties of surfactants as explained in CH1, and dynamic surface tension gives useful features of surfactants under the transient or non-equilibrium conditions such as their concentration dependency or detergent foaming, both of which are key factors for industrial application. For example, adsorption of surfactants to interfaces causes a reduction in interfacial tension which correlates to emulsification or wetting performance.

In this chapter, maximum bubble pressure measurement is introduced as a typical method to determine dynamic surface tension as shown in Fig. 3.1. Calculation of surface tension is based on the Young-Laplace equation [1, 2] (Eq. 3.1) for the bubble shown in Fig. 3.1 (4).

T. Ogura (✉)
Research & Development Headquarters, LION Corporation, Tokyo, Japan
e-mail: t-ogura@lion.co.jp

M. Abe (ed.), *Measurement Techniques and Practices of Colloid and Interface Phenomena*, https://doi.org/10.1007/978-981-13-5931-6_3

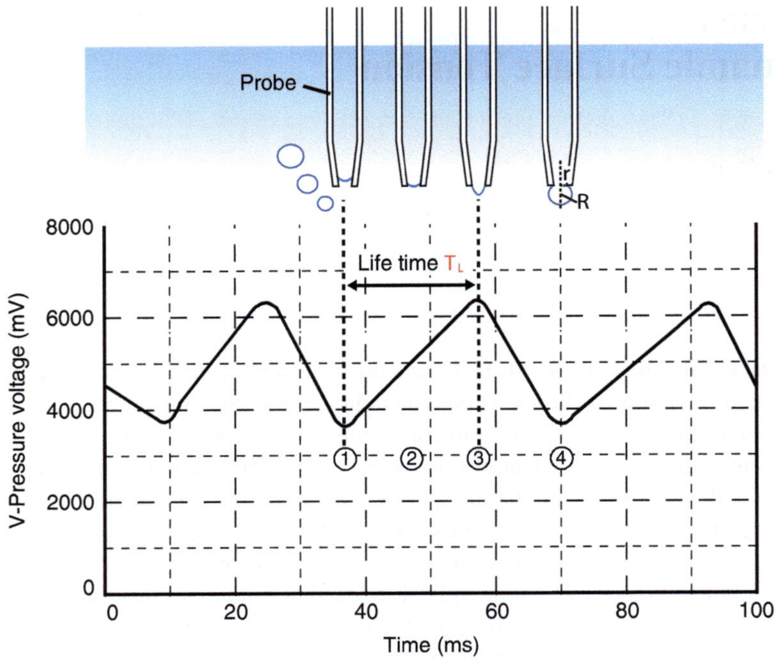

Fig. 3.1 Dynamic surface tension measurement by maximum bubble pressure method [3]

$$\Delta P = 2\gamma/R \tag{3.1}$$

where ΔP (Laplace pressure) is the pressure difference between the inside and the outside of the curved surface that forms the boundary between a gas region and a liquid region. The pressure difference is caused by the surface tension of the interface between liquid and gas. γ is the surface tension and R is the curvature radius of the bubble.

The maximum bubble pressure technique is a method for determining the surface tension from the maximum pressure. In order to measure the surface tension, pressurized air is sent continuously through the capillary as shown in Fig. 3.1, and the pressure inside the capillary (P) changes periodically. When the radius r of the capillary tip and the radius of curvature R of the bubble are equal (Fig. 3.1(3)), the pressure becomes maximum. In the next step, the pressure drops rapidly by expansion of the bubbles. The (1)–(3) in Fig. 3.1 is called the lifetime of bubbles (T_L, the time duration from when the new interface generates at the capillary tip to when the bubble pressure reaches its maximum). Surfactants adsorbed during the lifetime will determine the surface tension [3], so this lifetime is an important parameter to evaluate the dynamic surface tension.

P (inner pressure of the bubble) is measured by the sensitive differential pressure gauge during the bubbling at the tip of the capillary, and then the dynamic surface tension is calculated by Eq. 3.2 as a function of buoyancy and dynamic surface tension [4].

$$\gamma_t = r(P - (h + 2/3r)\rho g)/2 \quad (3.2)$$

where γ_t is the surface tension at time *t*, *r* is the inner radius of the capillary, *P* is the inner pressure of the bubble, h is the distance between the solution surface and the capillary tip, ρ is the density of the sample solution, and g is the acceleration of gravity.

3.2 What You Get

2.1. A calculated rate of the surface tension reduction ($d\gamma_t$/dt) max as a characteristic index of the surfactant
2.2. A calculated lifetime of the bubble (T_L) as an adsorption parameter of the surfactant
2.3. Prediction of the wettability in a short period of time

3.3 Essentials and Tips

Figure 3.2 shows a dynamic surface tensiometer BP-D5 (Kyowa Interface Science Co., Ltd.), which is based on the maximum bubble pressure method. This method is very sensitive to the conditions of the solution such as temperature, concentration, or contamination of any alien materials. The purity of the water and the instrument is

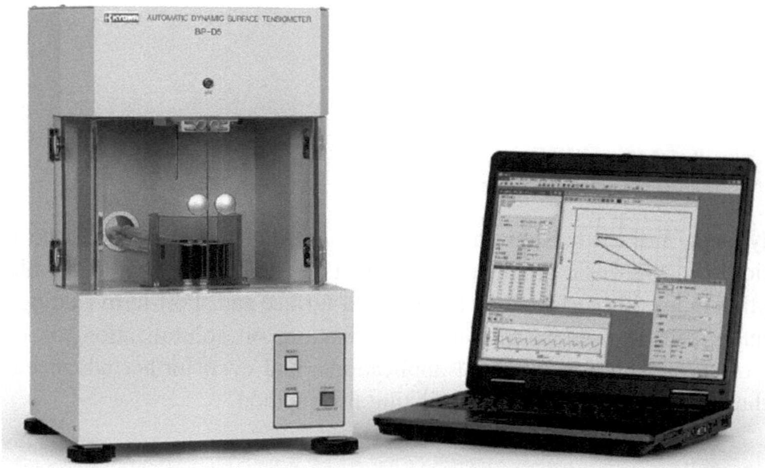

Fig. 3.2 Dynamic surface tensiometer BP-D5. (Photo credit Kyowa Interface Science Co.)

key to accurate measurement, and an ultrasonic cleaner for capillary cleaning is installed for BP-D5.

3.4 Understanding Your Data

The maximum surface tension reduction rate, as an index to show foaming power, is calculated by the Rosen fitting technique as shown in Eq. 3.3 [5].

$$\gamma_t = \gamma_m - (\gamma_0 - \gamma_m)/\left(1 + \left(1 + (t/t^*)^n\right)\right) \tag{3.3}$$

where t is the time (sec) of adsorption, γ_t is the surface tension at t, γ_m (mNm^{-1}) is the meso-equilibrium surface tension (time variation is less than 1 mNm^{-1}/30 s), γ_0 is the surface tension of the solvent (72.75 mNm^{-1} at 20 °C), and t* is the adsorption time when γ_t reaches the middle of γ_O and γ_m as shown in Eq. 3.4.

$$\gamma_{t \Rightarrow t*} = (\gamma_0 + \gamma_m)/2 \tag{3.4}$$

where n is a constant. A different form of Eq. 3.4 can be shown in Eq. 3.5.

$$d\gamma_t/\mathrm{dt} = (\gamma_0 + \gamma_m)(n(t/t^*)/t^*)/\left(1 + (t/t^*)^n\right)^2 \tag{3.5}$$

when $t = t^*$

$$d\gamma_t/\mathrm{dt} = -n(\gamma_0 + \gamma_m)/4t^* = -(d\gamma_t/\mathrm{dt})_{\max} \tag{3.6}$$

The maximum rate of surface tension reduction $(d\gamma_t/\mathrm{dt})_{\max}$ can be calculated from Eq. 3.6.

3.5 Useful Hints

The dynamic surface tension can be used to predict the rate of surface tension reduction as an intrinsic property of a surfactant and the lifetime of bubble as an adsorption index of a surfactant to a gas-liquid surface and short-term wetting. These parameters can be used to evaluate micellization, oil solubilization, or foaming power of surfactants for industrial applications, and even further advancement or breakthroughs can be expected by utilizing this method.

References

1. H.J. Butt, K. Graf, M. Kappl, *Physics and Chemistry of Interfaces* (Wiley-VCH, Berlin, 2006)
2. J.T. Davies, E.K. Rideal, *Interfacial Phenomena* (Academic, New York, 1961)
3. Kyowa Interface Science Co., Ltd., *"What is Surface Tension?"* http://www.face-kyowa.co.jp/english/en_science/en_theory/en_what_Surface_tension/
4. J.B.M. Hudales, H.N. Stein, *J. Colloid Interface Sci.* **140**, 307 (1990)
5. Y. Hua, M.J. Rosen, *J. Colloid Interface Sci.* **124**, 652 (1988); Y. Hua, M.J. Rosen, *J. Colloid Interface Sci.* **139**, 397 (1988); Y. Hua, M.J. Rosen, *J. Colloid Interface Sci.* **141**, 180 (1988)

Chapter 4
Surface Pressure

Tomohiro Imura

Abstract The basic concept of the Langmuir–Blodgett trough by measuring the surface pressure of monolayers on water was developed by Irving Langmuir. With this surface pressure measurement, we can evaluate the molecular orientations and interactions of lipids, oils, proteins, and surfactants spreading over liquids as a monolayer. Explorations of the functions of biomembranes utilizing interactions between specific lipid molecules and bioactive molecules at air/water interfaces are research targets based on this technology and are closely related to the advancement of the basics and applications for life science and medical fields. Moreover, the Langmuir–Blodgett film (LB film) as accumulated monolayers on the solid substrate is expected to be a useful module to develop molecular memory or molecular electronic devices.

Keywords Surface pressure · Area per molecule · Monolayers · Langmuir–Blodgett film

4.1 Introduction

The history of research on monolayer surfaces dates back to Benjamin Franklin's experiment in 1774. Franklin poured just a spoonful of olive oil onto the Clapham pond in England, which spread over as a thin film and seized the wave. Later in the 1880s, Agnes Pockels conducted a surface tension measurement in her kitchen by attaching a floating button to a balance, and the results were published in *Nature* in 1891 from the request of Rayleigh. In 1917, Irving Langmuir introduced the basic concept of the Langmuir–Blodgett trough by measuring the surface pressure of monolayers on water. With this surface pressure measurement, we can evaluate the molecular orientations and interactions of lipids, oils, proteins, and surfactants

T. Imura (✉)
Research Institute for Chemical Process Technology, National Institute of Advanced Industrial Science and Technology (AIST), Tokyo, Japan
e-mail: t-imura@aist.go.jp

M. Abe (ed.), *Measurement Techniques and Practices of Colloid and Interface Phenomena*, https://doi.org/10.1007/978-981-13-5931-6_4

spreading over liquids (mainly water) as a monolayer. Explorations of the functions of biomembranes utilizing interactions between specific lipid molecules and bioactive molecules at air/water interfaces, or development of biomimetic materials and bio-interface devices, are typical research targets based on this technology and are closely related to the advancement of the basics and applications for life science and medical fields. The Langmuir–Blodgett film (LB film) as accumulated monolayers on the solid substrate is expected to be a useful module to develop molecular memory or molecular electronic devices.

4.2 What You Get

1. Surface pressure (π) – the area per molecule (A) isotherm as a two-dimensional state equation
2. Investigation of molecular orientations and interactions of monolayers at air/water interfaces
3. Evaluation of molecular interactions and miscibility in binary systems
4. The Langmuir–Blodgett film (LB film) accumulated on the solid substrate

4.3 Essentials and Tips

There are two types of surface pressure devices. One is the Wilhelmy type with a platinum or paper plate vertically hung onto the water surface to measure the vertical force, and the other is the Langmuir type, with a Teflon-coated float shorter than the trough width that moves on the water to measure the horizontal force. The Wilhelmy plate is a rather common device due to its simpler structure and easiness to handle, and this type of instrument will be explained here in this chapter. Figure 4.1 shows a

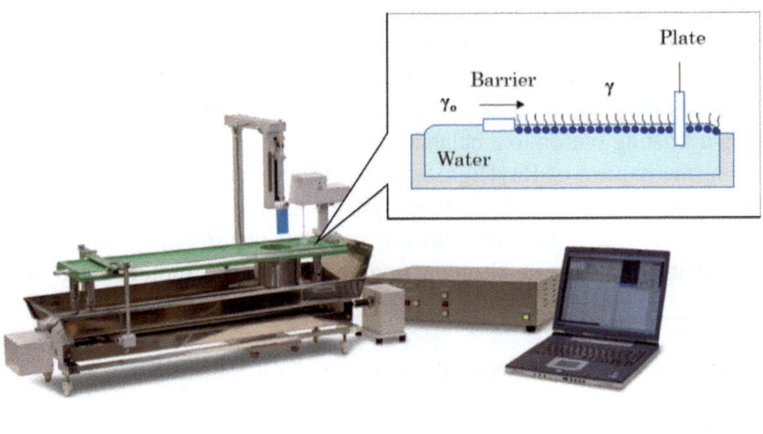

Fig. 4.1 Surface pressure meter (HBM-700. Photo credit Kyowa Interface Science Co. Ltd.)

Wilhelmy-type surface pressure instrument (HBM-700, Kyowa Interface Science Co., Ltd) with Langmuir–Blodgett trough. In the measurement procedure, first fill water over the edge of the clean trough, then prepare clean water surface, and level it (0.5–1 mm) by sweeping excess water off with a clean partition plate (barrier) or by sucking water out with an aspirator. After adjusting the trough temperature with a temperature-controlled bath, immerse the platinum plate after red heating with an alcohol lamp for cleansing, and then cool to room temperature. Next, calmly but rapidly drop the sample solution in the volatile solvent, such as chloroform, with a micro syringe. Keep still for over 10 min. to make the monolayer at the air/water interface from solvent evaporation.

The monolayer generates a spreading force at the air/water interface from the compression of the surface area with the partition plate (barrier). This force is called surface pressure (π), which is expressed by the difference of surface tension of pure water (γ_0) and the surface tension of the monolayer (γ) as $\pi = \gamma_0 - \gamma$.

Thus, π (mN/m) can be obtained by measuring the surface tension of the monolayer (γ).

4.4 Understanding Your Data

A surface pressure (π)–area per molecule (A) isotherm can be plotted using this instrument as shown in Fig. 4.2, where the monolayer area per molecule (A) is calculated by dividing the surface area by the number of molecules consisting the monolayer. This isotherm corresponds to the pressure (p) and volume (v) isotherm, so the (π)–(A) isotherm is regarded as a two-dimensional state equation, and the molecular interactions and orientations or miscibility of molecules composing the monolayer can be obtained.

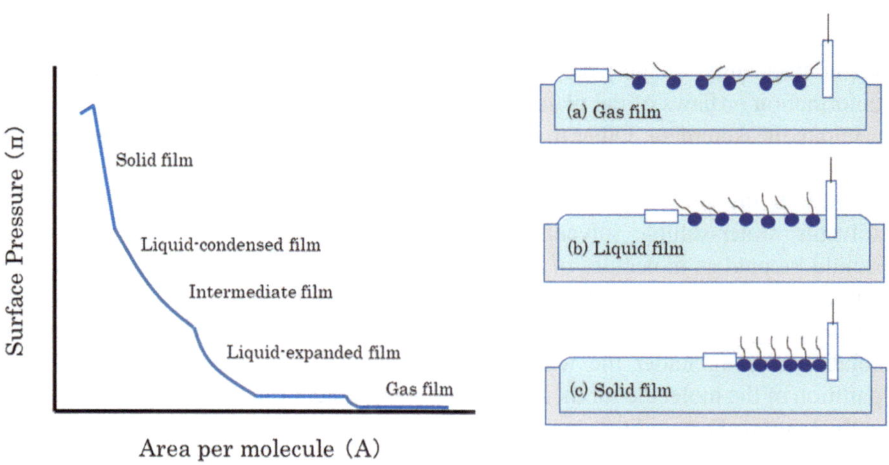

Fig. 4.2 Surface pressure (π)–area per molecule (A) isotherm

The profile of (π)–(A) isotherm (Fig. 4.2) shows the state condition of the monolayer. When the monolayer is not compressed by a partition plate (barrier), the molecules are free to move on the water surface; thus this region is called a gas film, as shown in Fig. 4.2a where the surface pressure (π) is small enough to be negligible. By compressing with a partition plate (barrier), the isotherm passes through a gas/liquid coexisting range where π stays constant regardless of the change in molecular area (A) and then reaches the liquid-expanded film where π increases compression while decreasing A. Further compression gives a liquid-condensed film through the intermediate film. Under these liquid film (Fig. 4.2b) regions, molecules are as mobile as liquids. With further compression, the two-dimensional state changes to solid film (Fig. 4.2c) and eventually reaches a collapsing pressure (γ_c) with the loss of its three-dimensional structure. As such, the (π)–(A) isotherm can be seen as a two-dimensional state equation. Its characteristics have a liquid expansion and intermediate regions (Fig.4.2) that do not exist in a three-dimensional state.

The surface compression ratio (β^s) can be calculated using Eq. 4.1 and the (π)–(A) isotherm, and the reciprocal number of β^s gives the Gibbs's modulus of the elasticity of the monolayer.

$$\beta^s = -\frac{1}{A}\left(\frac{\partial A}{\partial \pi}\right)_T \tag{4.1}$$

Here, A is molecular area, π is surface pressure and T is temperature.

As β^s is a numerical differentiation from the (π)–(A) isotherm, the phase transition of the monolayer can be accurately obtained by plotting β^s *vs* A.

4.5 What to Look Out for

Surface tension measurement using the Wilhelmy plate is based on the premise that the contact angle of the plate to water is zero. Therefore, any contamination and deformation or flaws of the plate that can affect the wettability would interfere with accurate measurement. Other than chloroform, organic solvents such as hexane and ethyl ether are commonly used to spread the monolayer molecule onto the water surface. If the solubility of the monolayer molecule is not sufficient with these solvents, water-soluble solvents such as ethyl alcohol can be used, but caution should be paid as molecules of the monolayer can dissolve into the lower aqueous phase and can interrupt accurate molecular area measurement. Other care is needed when using chloroform, whose density is larger than water due to chloroform solution sinking under the water, and can prevent solvent evaporation and the partition of the molecule for the monolayer in the water or in the chloroform solution at the bottom. To avoid this, put the tip of a micro syringe as close as possible to the water surface to spread the solution on the surface.

As the (π)–(A) isotherm is intended for insoluble monolayers, this method is not ideal for water-soluble materials such as hydrophilic surfactants.

4.6 Useful Hints

Monolayers at the air/water interface can be collected on to the solid surface, such as mica under constant surface pressure (π). This results in a film called the Langmuir–Blodgett film (LB film), which has been gaining attention as a potential molecular device with its nonconductive, piezo electric, and nonlinear optical properties. Atomic force microscope (AFM) observation is applicable by fixing the film on the solid base, and information relating to the orientation and miscibility of the monolayer at the molecular level can be obtained. Caution should be paid since the LB film does not always directly represent the condition of monolayer at the air/water interface. There are other instrumental methods to investigate monolayers at the air/water interface, such as the Brewster angle microscope (BAM), although its resolution is lower than AFM.

Chapter 5
Surface Viscosity

Tomohiro Imura

Abstract Foam stability is closely related to the liquid's surface viscosity under non-equilibrium unstable conditions. Surface viscosity measurements for surfactants, lipids, or protein solutions give reliable indices for the foam stability such as when stable soap bubbles can be made by adding thickening agents to the soap solution. There are mainly three methods to measure surface viscosity of aqueous solutions with small amounts of surface active materials such as surfactants or proteins: the damped oscillation, rotation, and capillary methods. In this chapter, damped oscillation is focused on as it is the most common method. Foaming is a key process for industrial products such as foamed plastics and cement, while anti-foaming is also important for the engineering process of food, fermentation, textiles, dyes, paints, ink, and petrochemical production.

Keywords Surface viscosity · Foam stability · Damped oscillation method

5.1 Introduction

Foaming is a common phenomenon in our daily life, like how soap wraps up dirt due to its huge surface area or how beer foams provide antioxidation and a smooth taste. The ease of foaming is attained by less work to enlarge gas/liquid interfaces and is closely related to the surface tension of the liquid. On the other hand, foam stability is closely related to the liquid's surface viscosity under non-equilibrium unstable conditions. Surface viscosity measurements for surfactants, lipids, or protein solutions give reliable indices for the foam stability such as when stable soap bubbles can be made by adding thickening agents to the soap solution.

Foaming is a key process for industrial products such as foamed plastics and cement, while anti-foaming is also important for the engineering process of food, fermentation,

T. Imura (✉)
Research Institute for Chemical Process Technology, National Institute of Advanced Industrial Science and Technology (AIST), Tokyo, Japan
e-mail: t-imura@aist.go.jp

M. Abe (ed.), *Measurement Techniques and Practices of Colloid and Interface Phenomena*, https://doi.org/10.1007/978-981-13-5931-6_5

textiles, dyes, paints, ink, and petrochemical production. The surface viscosity gives important pointers for the design and selection of anti-foaming agents.

5.2 What You Get

1. The surface viscosity for surfactants, lipids, or proteins in various solutions. Water is the most typical solvent in this application
2. The time dependence of surface viscosity, which is an important index of tracing foam stability
3. Information of molecular orientation and mobility of adsorbed layers as molecular self-assembly
4. Interfacial viscosity at liquid/liquid interfaces as an index of emulsion stability

5.3 Essentials and Tips

There are three methods to measure surface viscosity of solutions of water with small amounts of surface active materials such as surfactants or proteins: the damped oscillation, rotation, and capillary methods. The damped oscillation apparatus consists of a flat disc-shaped detector horizontally fixed to the shaft and a table to put the sample solution in the dish. Rotational oscillation is applied to the disc after placing it on top of the sample surface, and the surface viscosity can be measured from the decay rate of rotational oscillation from the viscosity resistance of the contacted liquid surface. The rotation method and capillary method are two-dimensional modifications of the Couette viscometer and capillary viscometer, respectively. In this chapter, damped oscillation is focused on as it is the most common method.

The surface viscometer SVR-A made by Kyowa Interface Science Co., Ltd. is shown in Fig. 5.1. In the measurement procedure, set the clean dish in the instrument's temperature-controlled chamber, fill up the sample solution over the edge of a

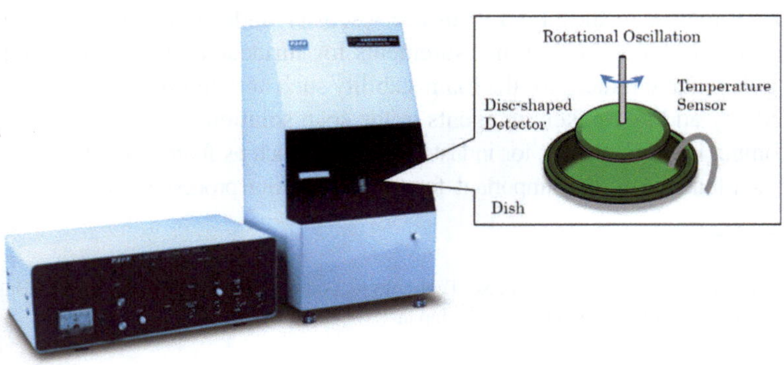

Fig. 5.1 Surface viscometer (type SVR-A. Photo credit Kyowa Interface Science Co. Ltd.)

clean dish, and then level the surface by sweeping excess sample solution off with a clean partition plate (barrier), or suck it out with an aspirator. Put the temperature sensor in the solution as surface viscosity is quite sensitive to temperature change. Move the table up so the detector contacts the solution surface, and then apply rotational oscillation to measure the surface viscosity from the decay rate of oscillation by the viscosity resistance of the contacted liquid surface.

5.4 Understanding Your Data

Data obtained by the rotational oscillator is the time dependency of oscillation displacement as a decay curve of rotational oscillation caused by surface viscosity as shown in Fig. 5.2. The decay rate (λ) is given by $\lambda = \log_e(a_1/a_2) = \log_e(a_2/a_3) = \cdot\cdot\cdot$, and the surface viscosity (η^s) can be calculated by Eq. 5.1.

$$\eta^s = \frac{I}{2\pi} \cdot \frac{\lambda}{T} \left(\frac{1}{\gamma_1{}^2} - \frac{1}{\gamma_2{}^2} \right) \tag{5.1}$$

where I is moment inertia, λ is natural logarithmic decrement, T is oscillation period, γ_1 is disc radius, and γ_2 is dish radius. The unit of the resulted two-dimensional surface viscosity is ($N/m\cdot\ S$), while the unit of viscosity at three dimension is (Pa$\cdot$ $S = N/m^2\cdot\ S$).

The surface viscosity of water at 25 °C is 0.10 mN/$m\cdot\ S$, which decreases with temperature increase due to the weakening of intermolecular hydrogen bonds. Sodium dodecyl sulfate solution (SDS; 50 mM at 25 °C), as one of the most typical anionic surfactants, shows 0.11 mN/$m\cdot\ S$, while the surface viscosity of commercial beer is c.a. 0.16 mN/$m\cdot\ S$ at 25 °C. As such, there is common relationship between surface viscosity and foam stability.

As explained above, water has a high surface viscosity due to its strong molecular interactions. Therefore, blank correction of water contribution is required to evaluate the actual surface viscosity of the adsorbed layer of the surfactant or insoluble lipid

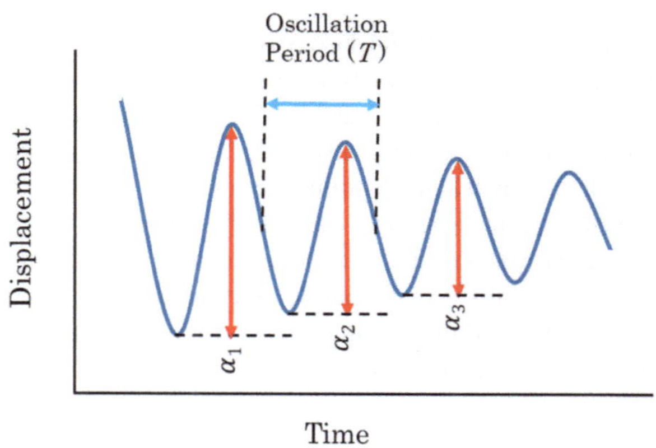

Fig. 5.2 Decay curve of rotational oscillation

monolayer at the water surface. This corrected surface viscosity can be calculated by Eq. 5.2.

$$\eta_S = \frac{I}{2\pi}\left(\frac{\lambda}{T} - \frac{\lambda_w}{T_w}\right)\left(\frac{1}{\gamma_1{}^2} - \frac{1}{\gamma_2{}^2}\right) \tag{5.2}$$

where I is moment inertia, λ is apparent natural logarithmic decrement observed, λ_w is natural logarithmic decrement of water, T is apparent oscillation period observed, T_w is oscillation period of water, γ_1 is disc radius, and γ_2 is dish radius. From this calibration, the surface viscosity of the actual adsorbed layer can be measured, and the molecular orientation and self-assembled states can be obtained.

5.5 Useful Hints

There are some useful hints for the measurement based on the damped oscillation method using surface viscometer SVR-A. Although the height of the sample solution in the dish does not need to be very sensitive for the measurement, it is desirable to keep the level consistent and preferably be equal to the edge of dish throughout the measurement. The contact position of the disc and sample surface significantly affects the results, and careful attention should be paid. The surface viscometer SVR-A is designed to control this positioning automatically by detecting the flow of electric current when the disc contacts the surface and stops the table elevation. (The actual stopping position is c.a. 0.5 mm above the point of contact.) As shown in Fig. 5.1, the temperature sensor works as a lead to detect electric current, so the sensor must be put in the sample solution prior to the measurement. With non-conductive solvents other than water, the table movement must be stopped manually when the sample contacts the surface with careful side view observation.

5.6 What to Look Out for

Liquid/liquid interfacial viscosity such as water and oil interface can be measured using the damped oscillation method with surface viscometer SVR-A in addition to the surface viscosity of gas/liquid interface (surface). For this measurement, put the surfactant solution in the dish with c.a. 5 mm depth, then move the table upward, and stop manually when the point disc contacts the solution surface. Gently pour oil up to the edge dish at level, and the disc should be completely in the oil phase at that point. This measurement can be done in the exact same manner with gas/liquid sample, and the interfacial viscosity can be determined by the damped oscillation method. It is well recognized that the interfacial viscosity obtained here shows a good correlation to the emulsion stability.

Chapter 6
Interfacial Tension Between Water and Oil

Takeshi Misono

Abstract There are many products (such as cosmetics, food, etc.) in our daily life made of emulsions, mixtures of water and oil. However, water and oil are essentially immiscible, and both liquids tend to separate into two phases. Interfacial tension is the force acting between two different liquids. Higher interfacial tension means that both liquids tend to separate into two phases. The higher the interfacial tension value, the more likely the phase separation of the two liquids occurs. Thus, decreasing the interfacial tension is essential to create a stable emulsion. As such, the interfacial tension is an indicator for better understanding stable emulsion formation. Two kinds of measurement methods, the Wilhelmy and pendant drop, are usually used for evaluating interfacial tension. The Wilhelmy method using a platinum plate is a method utilizing the drawing force of liquid to the plate, and this force is converted into interfacial tension. The pendant drop method is a method of applying the curvature radius of the liquid and is often used by many researchers in recent years. The principle and detailed information of evaluating interfacial tension is described in this chapter.

Keywords Interfacial tension · Wilhelmy method · Pendant drop method · Emulsifier

6.1 Introduction

Liquid/liquid interfacial tension refers to the force contributing to the interface between two liquids (generally water and oil) which are essentially immiscible. Simply put, the principles of Chaps. 1 and 2 are applied but replacing air with liquid. However, there are many products (such as cosmetics, food, etc.) in our daily life made of emulsions, mixtures of water and oil. These emulsion products are produced by mixing water and oil and by adding emulsifiers such as surfactants. It is important

T. Misono (✉)
NIKKOL Group Cosmos Technical Center Co., Ltd., Tokyo, Japan
e-mail: t.misono@ns-cosmos.co.jp

M. Abe (ed.), *Measurement Techniques and Practices of Colloid and Interface Phenomena*, https://doi.org/10.1007/978-981-13-5931-6_6

to know the interfacial properties of an emulsifier to better understand the emulsion properties. Interfacial tension measurement is an effective indicator and will be described in this chapter.

6.2 What You Get

An understanding of the interfacial tension between different liquids and the properties of emulsifiers

6.3 Essentials and Tips

Oil-water interfacial tension measurement can be measured by methods with principles similar to surface tension methods. The two most popular methods are the Wilhelmy and pendant drop method.

6.3.1 Wilhelmy Method

When a platinum plate is immersed in the liquid solution, the wettable liquid spreads on the plate, and a drawing force of liquid occurs. The Wilhelmy method measures this force which corresponds to the interfacial tension. Figure 6.1 shows an equipment using the measurement principle shown in Fig. 6.2. In this principle, the interfacial tension is calculated by the following equation.

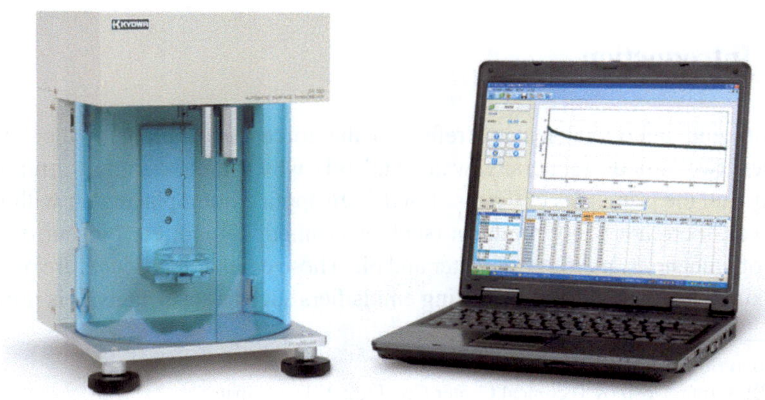

Fig. 6.1 Interfacial tensiometer (Photo credit Kyowa Interface Science Co., Ltd.)

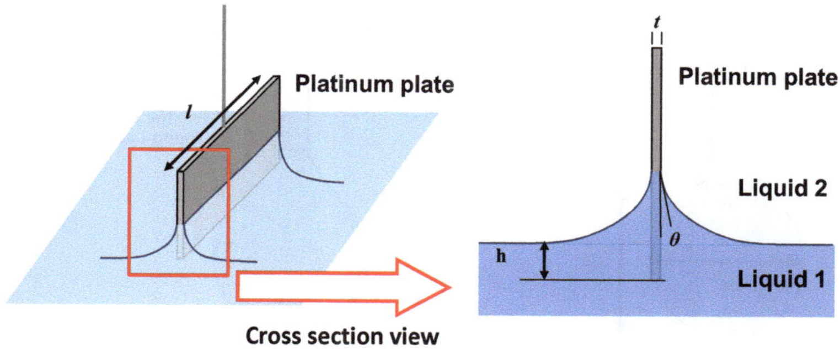

Fig. 6.2 Measurement principle of Wilhelmy method

$$P = \mathrm{mg} + 2(l + t)\gamma \cdot \cos\theta - \mathrm{lth}\Delta\rho g \tag{6.1}$$

P: tension
m: plate weight
g: gravity acceleration
l: length of plate
t: plate thickness
γ: interfacial tension
θ: contact angle between the plate and liquid
h: depth
Δρ: density difference of liquids

As seen in the equation above, The Wilhelmy method can measure the tension on the plate to calculate the interfacial tension. Therefore, it is necessary to use a hermetically sealed chamber as shown in Fig. 6.1 to avoid external factors such as wind, vibration, etc.

6.3.2 Pendant Drop Method

The pendant drop method is a method used to calculate the interfacial tension from the curvature radius of a droplet. The parameters for the calculation of the interfacial tension in the droplet are shown in Fig. 6.3. The relationship between the interfacial tension and each parameter can be expressed by the following equation:

$$\gamma = \Delta\rho \mathrm{gd}_e{}^2 \cdot \frac{1}{H} \tag{6.2}$$

γ: interfacial tension
Δρ: density difference of liquids
g: gravity acceleration

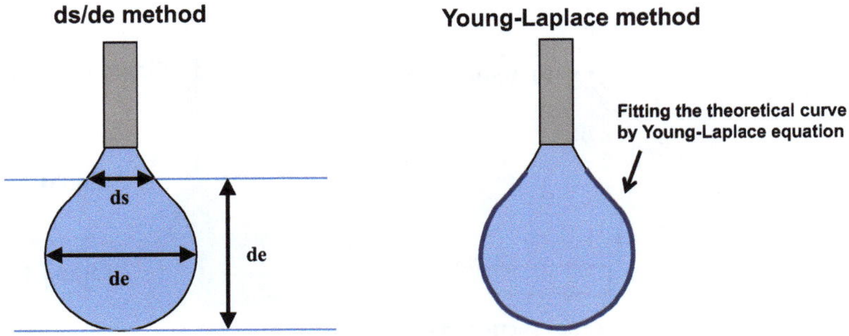

Fig. 6.3 Measurement principle of pendant drop method

d_e: maximum diameter of the droplet
1/H: correction coefficient obtained from d_s/d_e

This is called the ds/de method and is used to easily calculate the interfacial tension.

Additionally, recent development in image analysis technology has made the Young-Laplace method more popular, a method that can accurately calculate the interfacial tension by fitting the outline shape of the droplet to the Young-Laplace equation. The shape of the droplet contour can be calculated by the following simultaneous differential equation from the Young-Laplace principle.

$$\frac{\mathrm{dx}}{\mathrm{ds}} = \cos\phi, \frac{\mathrm{dz}}{\mathrm{ds}} = \sin\phi, \frac{d\phi}{\mathrm{ds}} = 2 + \beta z - \frac{\sin\phi}{x} \left(\text{where}, \beta = -\frac{\Delta\rho \mathrm{gb}^2}{\gamma}\right) \quad (6.3)$$

The interfacial tension can be estimated more accurately than the ds/de method by fitting the numerical plot on the contour curve and theoretical curve.

6.4 Understanding Your Data

Higher interfacial tension value indicates that the two liquids are difficult to mix, meaning both liquids tend to separate more easily into two phases. Emulsifiers such as surfactants are often used to avoid this phase separation. The interfacial tension is decreased by adsorption of the emulsifiers into the liquid/liquid interface and, as a result, prevents phase separation. The relationship between interfacial tension and emulsifier concentration is shown in Fig. 6.4. This shows that the interfacial tension decreases by increasing emulsifier concentration so that energy required for emulsification, namely, creation of new interfaces, can be reduced. The emulsion size also depends on the interfacial tension. For example, emulsions can be prepared with normal stirring when the interfacial tension is lower than 30 mN/m, and emulsions can be prepared spontaneously (without stirring) when the interfacial tension is

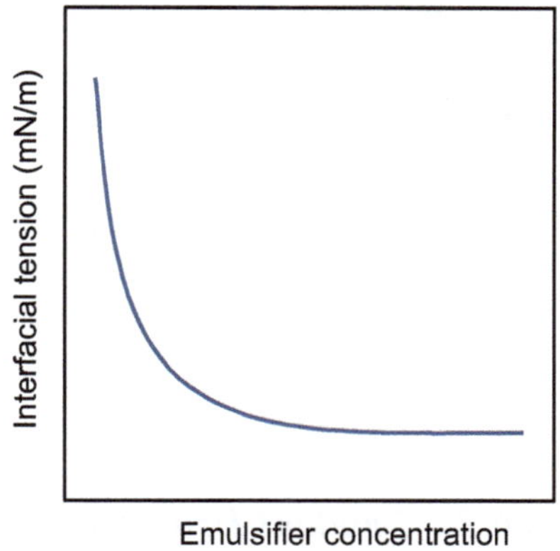

Fig. 6.4 Plots of interfacial tension as a function of emulsifier concentration

lower than approx. 2 mN/m. Furthermore, when the interfacial tension is reduced to 10^{-3} mN/m, a very fine emulsion (nano-emulsion) can be prepared.

6.5 What to Look Out for

6.5.1 Wilhelmy Method

It is common to conduct the measurement where the contact angle value θ is set to zero to avoid the influence of that angle. The contact angle θ can be set to zero by selecting the plate with a well wettable sample solution and plate surface treatment.

6.5.2 Pendant Drop Method

It is necessary to pay attention to the formation of droplets in order to calculate the interfacial tension from the droplet shape. In general, it is considered appropriate to keep de/ds ratio from 0.6 to 0.9. Furthermore, this method is not suitable for long-time measurement since the droplet curvature depends on the droplet amount.

It is necessary to fill the syringe with only the sample solution to keep the internal pressure of the syringe constant. For example, if air is contaminated in the syringe, it is impossible to keep the droplet amount constant by the change of internal pressure. It is recommended that the syringe containing the sample solution is turned upside down and the inside air is pushed out before setting it to the measuring equipment.

6.6 Useful Hints

6.6.1 Wilhelmy Method

In this method the plate is set at the interface for measurement. If either the aqueous or the oil phase is colored with a small amount of dye, the interface can be visually confirmed. In the initial setting stage of measurement, the wettability to the plate can be easily determined by this treatment.

6.6.2 Pendant Drop Method

Continuous measurement for interfacial tension against emulsifier concentration is possible with this method by successively adding emulsifier solution to the continuous phase.

Chapter 7
Quartz Crystal Microbalance with Dissipation Monitoring (QCM-D)

Kenichi Sakai

Abstract Quartz crystal microbalance with dissipation monitoring (QCM-D) can detect changes in oscillating frequency of a quartz crystal as well as in energy dissipation of adsorption material. For example, a decreased frequency results from an increased adsorption amount, while an increased dissipation results from an increased viscoelasticity of the adsorption layer. The QCM-D can monitor the frequency and dissipation changes with less than 1 s intervals. This enables us to analyse non-equilibrium adsorption or reaction system in solution. Various sensors such as gold, silica, titania, iron oxide or polystyrene are available for the QCM-D instrument, which enables QCM-D measurement possible for a wide range of applications. Special attention should be paid to the solvation effect of the adsorption material; the amount obtained by QCM-D is expressed as "wet mass", while that of other methods such as ellipsometry or optical reflectometry is called "dry mass". Another important point in QCM-D measurement is the effect of viscosity and density of bulk solution because these properties also contribute to the frequency change.

Keywords Quartz crystal microbalance with dissipation monitoring (QCM-D) · Frequency · Energy dissipation · Adsorption

7.1 Introduction

Measuring adsorption isotherms is indispensable for evaluating the adsorption behaviour of surfactants at solid/liquid interfaces. Adsorption isotherm measurement has been practiced for many years, and the depletion method is one of the most common procedures. This method consists of adding surfactants into a dispersed solid particle system and then measuring the equilibrium concentration of the

K. Sakai (✉)
Department of Pure and Applied Chemistry, Faculty of Science and Technology, Tokyo University of Science, Tokyo, Japan
e-mail: k-sakai@rs.noda.tus.ac.jp

M. Abe (ed.), *Measurement Techniques and Practices of Colloid and Interface Phenomena*, https://doi.org/10.1007/978-981-13-5931-6_7

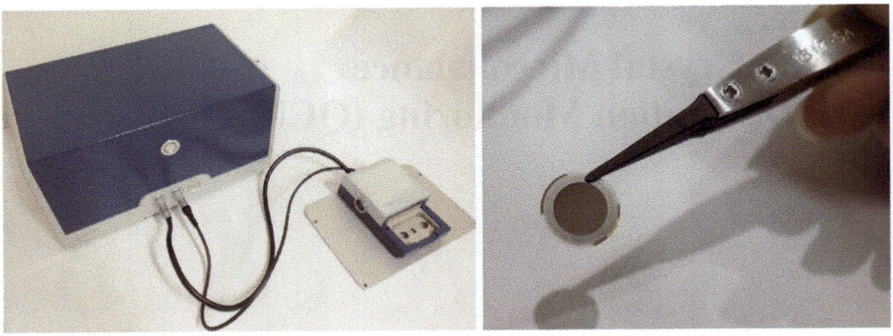

Fig. 7.1 Q-Sense QCM-D instrument

surfactant in the bulk solution after a certain time period. The adsorption amount per unit mass or area of the solid particle can be calculated by the concentration difference of the surfactant before and after adsorption. Although requiring multiple steps is a drawback, the depletion method has advantages for the adsorption isotherm measurement as it requires only common analytical instruments and procedures such as high-performance liquid chromatography (HPLC) analysis or solvent extraction of surfactant complex formed with dye.

In recent years new methods of evaluating adsorption onto flat solid surfaces have been developed, such as quartz crystal microbalance with dissipation monitoring (QCM-D), optical reflectometry (OR), ellipsometry and surface plasmon resonance (SPR) techniques. Although these new instrumentations are rather expensive and not easily available at the time needed, useful information can be obtained with the appropriate knowledge of their advantages and limitations as well as their well-designed experimental preparation. The principle of QCM-D is to measure the change (reduction) in oscillating frequency of a quartz crystal corresponding to the adsorption amount of material based on the Sauerbrey equation (Eq. 7.1) [1]. Energy dissipation can also be measured (Eq. 7.2) along with the frequency change oscillator using QCM-D (Fig. 7.1) by Q-Sense corporation. The amplitude of oscillation decays by cutting off the electric current for the oscillator. Abrupt decay of amplitude will be found for adsorbed soft viscoelastic materials, while hard adsorbed materials with lower viscoelasticity keep oscillation with slow decay. Namely, dissipation is a useful indicator to represent rheological properties of the adsorbed material, i.e. higher viscoelasticity is larger in dissipation (Fig. 7.2) [2]. As such, QCM-D is a highly sensitive instrument capable of detecting tiny amounts of adsorption, but it should be noted that results could be easily interfered due to additional factors. Those points will be discussed further in the following sections based on the author's experience with Q-Sense QCM-D.

$$\Delta m = -C\frac{\Delta f_n}{n} \tag{7.1}$$

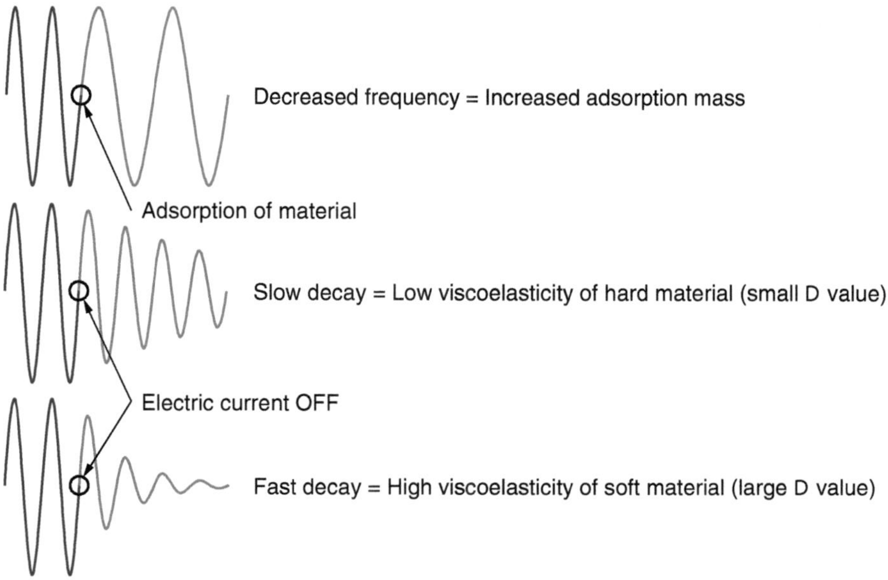

Fig. 7.2 Schematic representation of QCM-D measurements: frequency change and dissipation change

Δm: Mass change
Δf_n: Frequency change observed at the overtone number n
n: Overtone number
C: Constant

$$D = \frac{E_{\text{lost}}}{2\pi E_{\text{stored}}} \tag{7.2}$$

E_{lost}: Lost or dissipated energy
E_{stored}: Stored energy

7.2 What You Get

1. A highly sensitive detection of adsorption amount, e.g. 1 Hz deviation in the resonance frequency for a quartz oscillator with a 5 MHz fundamental frequency corresponds to 17.7 ng/cm^2 = 0.177 mg/m^2 of the mass change.
2. The viscoelastic property of the adsorbed layer can be detected by monitoring energy dissipation, along with the adsorption mass.
3. Adsorption and reaction in both solutions and in gases, along with frequency and dissipation changes in 0.3–0.5 s intervals, can be analysed. This means that non-equilibrium analysis can also be performed as shown in Figs. 7.3 and 7.4, which is difficult to conduct by the depletion method.

Fig. 7.3 Schematic representation of QCM-D data

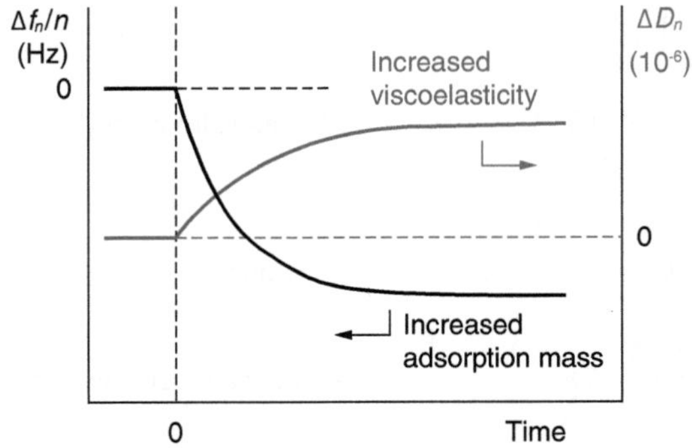

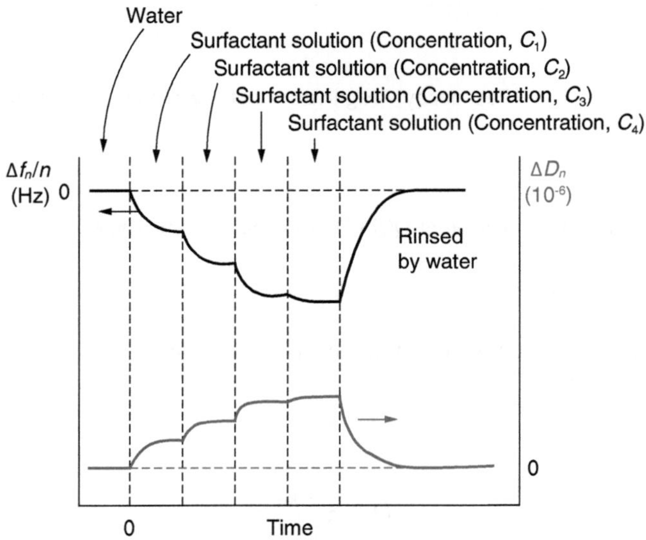

Fig. 7.4 Schematic representation of adsorption isotherm data through QCM-D measurements

4. Viscoelasticity and thickness of adsorbed layer can be estimated based on the rheological modelling by utilizing attached analytical software provided with the instrument.

7.3 Essentials and Tips

1. External vibration such as opening and closing of the door and walking alongside or nearby instrument could be a source of noise, so the instrument placement should be chosen carefully.

2. Frequency and dissipation changes using QCM-D measurement are very sensitive to temperature change. For the Q-Sense QCM-D, temperature control is set to be ±0.02 °C. Temperature monitoring and recording is useful to investigate the data with noise.
3. Sensor surface cleansing is an inviolable rule to obtain accurate data, and the cleansing method is different depending on the sensor material. For silica sensors, UV-ozone cleansing followed by repeated ultrasonic cleansing in weakly alkaline-surfactant solution and rinsing by pouring pure water and then drying by nitrogen gas flow is a common practice to decompose and remove organic contaminants on the sensor. Clean tweezers should be used as it could be a source of contamination.
4. The resonance frequency and dissipation values in both air and pure water must be recorded before each measurement. This record is a useful indicator to monitor sensor deterioration.
5. Avoid formation and contamination of air bubbles during the measurement. Special care should be paid for the surfactant solution as its foamability. Degassing the sample solution is desirable for measurements at higher temperatures than an injection temperature into the cell to avoid bubble formation due to the solubility reduction of gas by temperature.
6. The cell module's O-ring should be changed regularly as its deterioration can cause distortion due to loss of elasticity, which can be a source of noise.

7.4 Understanding Your Data

By using QCM-D measurement, the amount of material adsorbed on the sensor can be estimated using the Sauerbrey equation. However, the differences in mass obtained by the measurement should not be confused with the actual adsorption amount, and special attention should be paid to the solvation effect. The estimated adsorption amount measured using QCM-D tends to be larger than that obtained by other methods such as OR, ellipsometry or SPR, and such differences between the results from QCM-D and other methods correspond to the solvation effect [3]. Therefore, the amount obtained by QCM-D is expressed as "wet mass", while that of other methods such as OR is called "dry mass". When the solvation is significant for the adsorption mass, the energy dissipation value also becomes large, corresponding to the increase in viscoelasticity of the adsorbed layer. There is an article reporting that the difference between wet and dry masses becomes insignificant under the consideration of surface roughness of the sensor [4].

Another important point in QCM-D measurement is the effect of viscosity and density of the bulk solution. Frequency will decrease from adsorption on the sensor surface (mass increase), but also increased bulk viscosity and density will reduce the frequency further. It is known that the degree of frequency reduction is in proportion to the square root of the product of bulk viscosity and density of the Newtonian fluid, as shown by the Kanazawa-Gordon equation [5] (Eq. 7.3). In order to evaluate the adsorption amount of surfactant using QCM-D measurement, solution viscosity and density are assumed to be constant under the experimental condition. For accurate frequency determination, predictable frequency change must be subtracted from the Kanazawa-Gordon equation, or effect of solution viscosity and density experimentally obtained with non-adsorbable sensor must be subtracted [6].

$$\Delta f = -f_0^{\frac{3}{2}} \left(\frac{\rho_L \eta_L}{\pi \rho \mu} \right)^{\frac{1}{2}} \tag{7.3}$$

Δf: Frequency change
ρ: Density of oscillator
ρ_L: Density of solution
f_0: Fundamental frequency
μ: Shear modulus of oscillator
η_L: Viscosity of solution

7.5 Useful Hints

Various sensors such as gold, silica, titania, alumina, hydroxyapatite, iron oxide or polystyrene are available for the Q-Sense QCM-D, which enables QCM-D measurement possible for a wide range of applications. A gold sensor is mandatory for the SPR method, and for OR and ellipsometry, the choice of sensors is limited since reflectivity of incident light is required for these methods.

References

1. G.Z. Sauerbrey, Physics **155**, 206 (1959)
2. M. Rodahl, F. Höök, A. Krozer, P. Brzezinski, B. Kasemo, Rev. Sci. Instrum. **66**, 3924 (1995)
3. F. Höök, B. Kasemo, T. Nylander, C. Fant, K. Sott, H. Elwing, Anal. Chem. **73**, 5796 (2001)
4. C. Gutig, B.P. Grady, A. Striolo, Langmuir **24**, 4806 (2008)
5. K.K. Kanazawa, J.G. Gordon II, Anal. Chim. Acta **175**, 99 (1985)
6. R. Bordes, F. Höök, Anal. Chem. **82**, 9116 (2010)

Chapter 8
Atomic Force Microscope (AFM)

Kenichi Sakai

Abstract Atomic force microscope (AFM) evaluates morphology of surfactant layer formed at a solid/liquid interface as well as additional information of the adsorption layer such as thickness and compressibility. The morphology can be visualized when a probe tip (cantilever) is scanned at a given repulsive set-point force arising from the adsorbed surfactant layer. This AFM observation method is called "soft contact method" due to its non-distractive nature to the soft materials. In force curve measurement, the separation "0" is defined as the constant compliance region where the cantilever and base substrate move together. Special care is required for the separation definition; if the adsorption layer is stacked like a sandwich between the cantilever and the base substrate, it would not be pushed out by compression. Spherical particles can be placed on the cantilever tip to measure the force and friction curves. This procedure is called "colloid probe method" and is useful in determining the interaction surface energy under the Derjaguin approximation.

Keywords Atomic force microscope (AFM) · Force curve · Adsorption

8.1 Introduction

Evaluation of amphiphilic molecule adsorption that occurs at solid/liquid interfaces has commonly been evaluated by measuring adsorption isotherm or zeta potential. Several adsorption models for the self-assembly of small amphiphilic molecules on the solid surface such as the reverse orientation model, surface bilayer model, and surface micelle model have been proposed [1]. In addition to these indirect methods, direct measurement by means of atomic force microscope (AFM; Fig. 8.1) was developed in the late 1990s. AFM observation can be performed using a sharp probe

K. Sakai (✉)
Department of Pure and Applied Chemistry, Faculty of Science and Technology, Tokyo University of Science, Tokyo, Japan
e-mail: k-sakai@rs.noda.tus.ac.jp

M. Abe (ed.), *Measurement Techniques and Practices of Colloid and Interface Phenomena*, https://doi.org/10.1007/978-981-13-5931-6_8

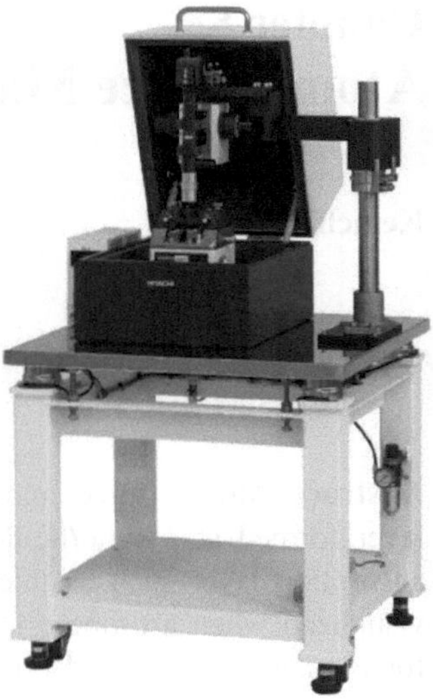

Fig. 8.1 AFM instrument. (Photo credit Hitachi High-Tech Science Corporation)

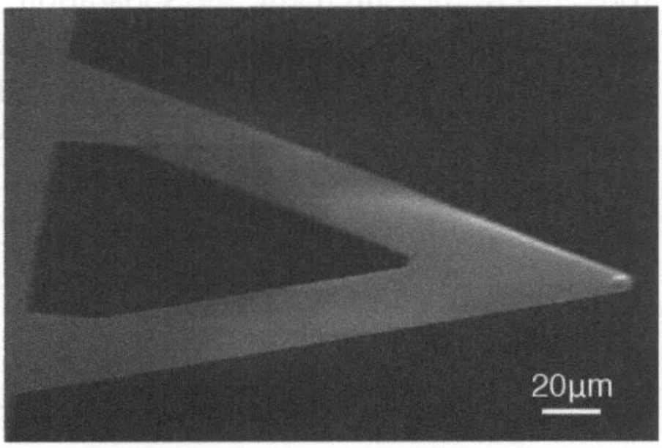

Fig. 8.2 SEM image of AFM cantilever

tip called a cantilever (Fig. 8.2), which can detect very weak repulsive force between the probe tip and molecular layers adsorbed on the solid surface in situ. Then the topographic information can be visualized when the cantilever is scanned at a given set-point force detected here [2]. This AFM observation method is classified as a type of contact method and is even called a soft contact method due to its non-distractive nature to the soft materials.

Additionally, the morphology of the adsorbed layers can be observed by nondestructive scanning of the cantilever under a certain oscillation. Structural evaluation of the adsorbed layers has previously been performed mostly in equilibrium, but now thanks to the recent technological advancement of AFM, we can observe and discuss the formation process of the adsorption layers [3, 4].

Forces between the cantilever and the adsorbed surfactant layers involve electrostatic interactions, steric repulsions, van der Waals attractions, hydrophobic interactions, etc. The soft contact method generally utilizes electrostatic interactions and steric repulsions. The force curve can be obtained by plotting the detected interactions vs. apparent surface separation. When the experimental condition results in attractive forces working between the probe and adsorbed layer, visualization of surface morphology is difficult in principle, even though valuable information for the adsorbed layer structure can be obtained from force curve measurements. The use of AFM allows us to get the visualized morphology of the adsorbed layer as well as further additional information through force curve interpretation.

As there are plenty of useful books available for the instrumental principles of AFM measurement, this chapter introduces some important points for the evaluation of adsorbed surfactants by AFM through the author's experiences.

8.2 What You Get

1. In situ evaluation for the morphology of the surfactant layer adsorbed on the solid surface such as spherical, globular, cylindrical (rod or thread), or planar (lamellar) morphology.
2. Information of adsorbed layers in depth direction such as the thickness and compressibility can be discussed from the force curve.
3. The dual force curve can be obtained by the movement direction of the cantilever; one is toward the substrate (approach), and the other is from it (retraction). The latter data give adhesive force of the cantilever to the adsorbed layer, but statistical processing is preferable in evaluating the attractive force since the fluctuation in data is quite high.
4. The friction curve (friction coefficient) can be measured for the adsorbed surfactant layer by measuring the twist amount (torsional response) under the constant vertical load on the cantilever. The friction coefficient as a nano-tribological parameter can be determined from the slope of the line plotted for the vertical load on the horizontal axis and the frictional force on the vertical axis.

8.3 Essentials and Tips

1. The main unit of AFM equipment should be placed on an anti-vibration table as external vibration would cause noise generation during the measurement.
2. The surface of the cantilever and the solid should be clean for the measurement, and the tweezers to hold the cantilever and cell for the sample solution should also be clean to avoid contamination.
3. The surface of the solid substrate should be as flat as possible to get accurate data and precise interpretation.
4. The thickness of the surfactant layer formed on the solid surface will be several nanometers, sometimes much smaller than the surface roughness of the solid substrate. As a result, the topographic image of the adsorbed layer based on the height profile becomes blurred (unclear). From this point, deflection or error signal image with enhanced edge morphology is usually used for the soft contact method.

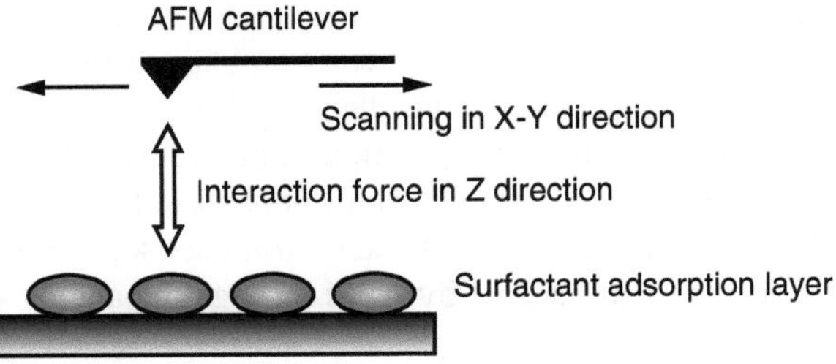

Fig. 8.3 Schematic representation of soft contact AFM method

5. For the soft contact method, the image of the adsorbed layer morphology is obtained by detecting weak repulsive forces between the cantilever tip and the adsorbed layer (Fig. 8.3). The adsorbed layer could be collapsed if excess force is applied to the adsorbed layer. If the applied force is too weak, the image would be too blurry since the cantilever scans the position far from the adsorbed layer. Hence, it is important to find the best set point to get an image with the highest resolution by measuring the force curve. Scanning at the point just before the cantilever jumps into contact gives the highest resolution since the force sensitivity to the distance becomes maximum. Caution should be paid for the thermal drift which causes shifting of the set-point force during the scanning, while this drift can be tactically utilized to observe morphology change of the adsorbed layer in response to the applied force. An example image for the surfactant layer with the soft contact AFM method is shown in Fig. 8.4.
6. It is important to change the area and direction of scanning when periodical image is found in the cylindrical structure in order to distinguish from electric noises.

8.4 Understanding Your Data

In situ observation and analysis are required to measure the adsorbed layers in solution, as its structure would change during the drying process. This should be sensibly kept in mind as we can find experimental reports trapped into this pitfall on the adsorbed structures of surfactants, especially polymer surfactants, observed under open atmosphere after evaporation of solvents.

AFM measurement gives force output change as electric voltage or current along with the piezo movement. By taking the spring constant and deflection amount of cantilever into account, the force curve can be obtained by plotting the detected interaction force against apparent surface separation (Fig. 8.5). Special caution should be paid here for the separation "0" is defined as the constant compliance region where the cantilever and base substrate move together. If the adsorbed layer is stacked like a sandwich between the cantilever and the base substrate, it would not be pushed out by compression, and the definition of separation "0" becomes obscure. On the other hand, when using surface force apparatus (SFA) [5], the distance

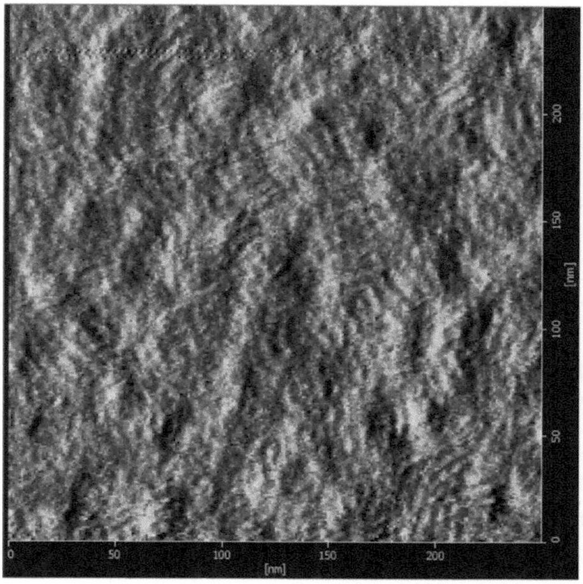

Fig. 8.4 Typical soft contact AFM image: adsorbed layer morphology of a cationic surfactant (hexadecyltrimethylammonium bromide, HTAB or CTAB) on silica in the presence of NaBr (250 nm image)

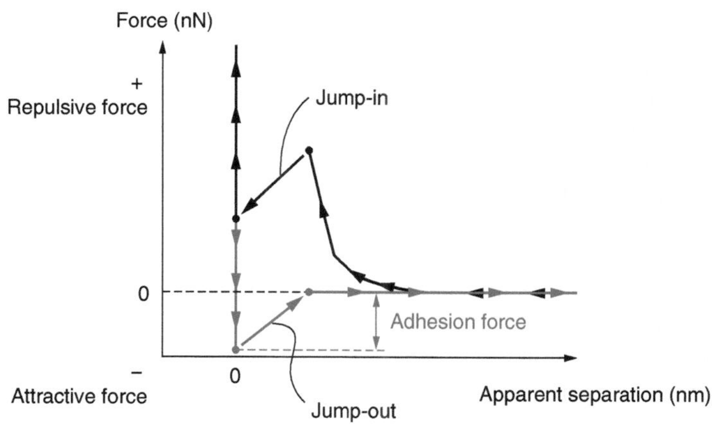

Fig. 8.5 Schematic representation of force curve data

between surfaces is strictly defined, and it causes a mismatch between the AFM and SFA data. Additionally, the interpretation of the spring constant can cause errors in force calculation.

Adsorption of surfactants onto the cantilever can be easily overlooked. Materials commonly used for cantilever are silicon or silicon nitride with negative charge in aqueous solutions under neutral pH. This causes adsorption of cationic surfactants on the cantilever and generates attractive force with the base substrate when the surfactant concentration is low while generating electrostatic repulsive force at high concentrations. The surfactant layer adsorbed on the cantilever is considered weak against compression, and two steps of repulsive force can be observed when the cantilever approaches toward the base substrate when the surfactant concentration is relatively high. The jump-in as a discrepancy of the first repulsion at longer distances and the second repulsion at shorter distances is interpreted as the adsorbed layer collapsing on the cantilever or base substrate [6]. The distance of the "jump-in" in the latter case corresponds to the thickness of the adsorbed layer [7], as shown in Fig. 8.5. There is another interpretation reported for the stepwise repulsion as derived from the double layers built up on the substrate surface [8]. It is recommended to apply other analytical methods such as ellipsometry for better understanding the phenomena.

8.5 Useful Hints

In relation to two to four of **What You Get**, spherical particles can be placed on the cantilever tip to measure the force and friction curves. This procedure is called the "colloid probe method" [9, 10]. Although common materials for the cantilever are silicon or silicon nitride, this spherical particle tip offers a wide choice of materials. Normalization using the curvature radius can convert the force data into the energy dimension with the Derjaguin approximation [5].

The curvature radius for the cantilever is usually provided from the manufacturer, but due to fluctuations between individual tips and unavoidable wearing, it is impossible to determine the accurate interaction energy from the provided radius. Furthermore, the Derjaguin approximation should be considered since it is applicable only when the distance of the surface force is negligibly smaller than the curvature radius of the tip [5]. Hence it is not desirable to determine the surface force (interaction energy) using the curvature radius of the tip. In contrast, for the colloid probe method, the curvature radius of the colloid probe can be measured, and its size is much greater than the force-detectable distance. From these points, the colloid probe method is useful in determining the interaction energy as a function of apparent surface separation.

References

1. K. Esumi, J. Colloid Interface Sci. **241**, 1 (2001)
2. S. Manne, J.P. Cleveland, H.E. Gaub, G.D. Stucky, P.K. Hansma, Langmuir **10**, 4409 (1994)
3. S. Inoue, T. Uchihashi, D. Yamamoto, T. Ando, Chem. Commun. **47**, 4974 (2011)
4. K. Koizumi, M. Akamatsu, K. Sakai, S. Sasaki, H. Sakai, Chem. Commun. **53**, 13172 (2017)
5. J.N. Israelachvili, *Intermolecular and Surface Forces*, 3rd edn. (Elsevier, San Diego, 2011)
6. F.P. Duval, R. Zana, G.G. Warr, Langmuir **22**, 1143 (2006)

7. E.J. Wanless, W.A. Ducker, J. Phys. Chem. **100**, 3207 (1996)
8. R.E. Lamont, W.A. Ducker, J. Am. Chem. Soc. **120**, 7602 (1998)
9. W.A. Ducker, T.J. Senden, R.M. Pashley, Nature **353**, 239 (1991)
10. W.A. Ducker, T.J. Senden, R.M. Pashley, Langmuir **8**, 1831 (1992)

Chapter 9
Static Light Scattering (SLS)

Takeshi Misono

Abstract Solid microparticles or nanoparticles and molecular aggregates formed by surfactants are dispersed in solutions in order of nanometers to submicrons. Static light scattering (SLS) is an excellent method that can nondestructively observe the aggregation number and detailed size information of these particles. Application of functions specific to nano- and microparticles has become active, and information of particles in the size and shapes mentioned above have important value in academic and industrial research. Various information can be obtained from SLS, for example, the molecular weight, the inertial radius, the second virial coefficient, and the aggregation number of colloidal particles in the solution. Additionally, information of the shape and size of particles can also be obtained. Analysis methods are classified as Debye plots and Zimm plots. To explain it more simply, small particles can be analyzed by Debye plots, while large particles are analyzed using Zimm plots. The principle and details of analysis methods by SLS measurement will be described in this chapter.

Keywords Static light scattering · Debye Plot · Zimm Plot · Particle size · Particle shape

9.1 Introduction

Nanometer- to submicron-order particles (molecular aggregates such as micelles, solid microparticles, etc.) are impossible to observe using normal microscopes. Light scattering is the most suitable method to know the structure of such particles. Static light scattering (SLS) measurement can give us the intensity of the scattered light as a function of scattering angle θ. As a result, the aggregation number, the inertial radius, the second virial coefficient, and other information can be estimated from the measurement. Recently, utilization of SLS is decreasing due to the widespread use of

T. Misono (✉)
NIKKOL Group Cosmos Technical Center Co., Ltd., Tokyo, Japan
e-mail: t.misono@ns-cosmos.co.jp

M. Abe (ed.), *Measurement Techniques and Practices of Colloid and Interface Phenomena*, https://doi.org/10.1007/978-981-13-5931-6_9

small-angle scattering (SAXS and SANS) devices. However, SLS is an excellent method that can nondestructively observe the aggregation number and detailed size information of the particles. In this section, analytical methods to estimate the aggregation number, size, and shape of colloidal particles by SLS are introduced.

9.2 What You Get

1. The molecular weight, inertial radius, second virial coefficient, and aggregation number of colloidal particles in the solution
2. Estimated shape and size of particles from the correlation using the theoretical curve

9.3 Essentials and Tips

There are several measurement methods for SLS, such as the Debye Plot method in which the measured angle is kept at 90°, and the Zimm Plot method in which the measured angle is changed. If the target particle size is smaller than 1/20 of the laser wavelength, the angular distribution of the scattered light intensity is symmetrical because of the angular dependency of the scattering intensity disappears. The angular dependency of this scattered light intensity can be examined from the value of the scattering light intensity ratio ($Z_{45} = I_{45}/I_{135}$). Herein, I_{45} and I_{135} are the scattering light intensity value when the scattering angle is at 45° and 135°, respectively. If this Z_{45} value is 1, there is no angle dependency, so the Debye Plot method can be used. On the other hand, if the Z_{45} value is larger than 1, an angle dependency exists so the Zimm Plot method must be used. Estimation of the particle shape and the aggregation number using the Debye Plot and the Zimm Plot method are described below.

9.3.1 Debye Plot Method

For small particles without scattering angle dependency, the following equation is obtained:

$$\frac{Kc}{R_\theta} = \frac{1}{M} + 2A_2c \tag{9.1}$$

K: Optical constant
c: Sample concentration [g/ml]
R_θ: Rayleigh ratio

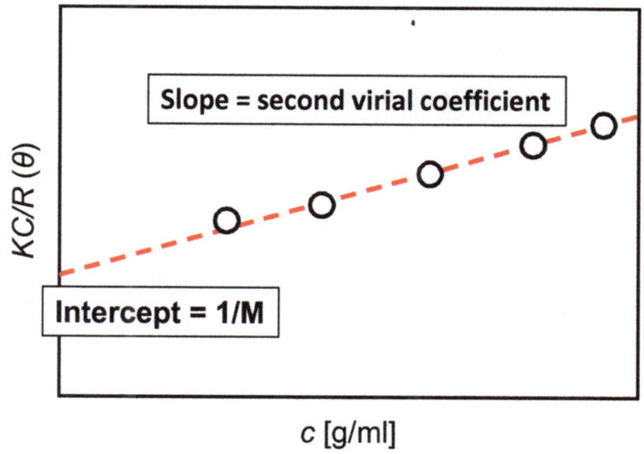

Fig. 9.1 Debye Plot method graph and the obtained information

M: Aggregation number (weight average molecular weight)
A_2: Second virial coefficient

As seen from Eq. 9.1, Debye Plot is shown by plotting Kc/R_θ as a function of sample concentration. Figure 9.1 shows the Debye Plot from which the second virial coefficient and weight average molecular weight can be estimated from the slope and intercept, respectively.

9.3.2 *Zimm Plot Method*

For large particles with scattering angle dependency, Eq. 9.1 can be transformed as Eq. 9.2:

$$\frac{Kc}{R_\theta} = \frac{1}{M}\left(1 + \frac{16\pi^2}{3\lambda^2}\langle R_g{}^2\rangle \sin^2\left(\frac{\theta}{2}\right) + \cdots\right) + 2A_2c \qquad (9.2)$$

λ: Laser wavelength [nm]
R_g: Inertial radius [nm]

The light scattering measurements at several angles are performed for various concentrations, and the Zimm Plot can be created by plotting "$Kc/R(\theta)$" in a function of "$\sin^2(\theta/2) + Bc$" as shown in Fig. 9.2. In this graph, the intercept represents the reciprocal of the weight average molecular weight. Then the aggregation number can be calculated by dividing the weight average molecular weight by the surfactant molecular weight. In addition, the second virial coefficient and the inertial radius can be estimated from the slope.

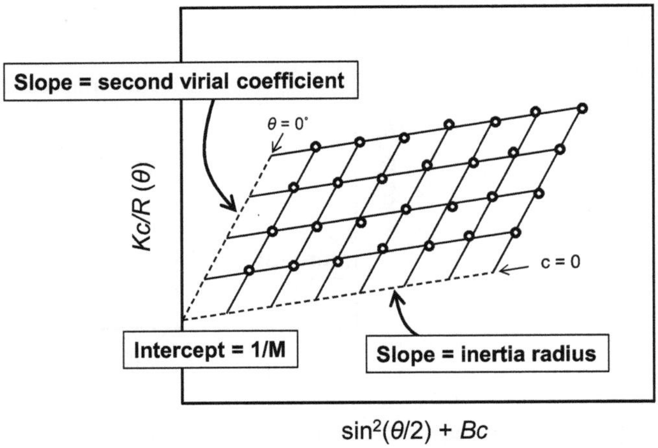

Fig. 9.2 Zimm Plot graph and the obtained information

9.3.3 *Estimation of Shape and Size*

The shape and size of the molecular aggregate can be estimated by comparing with the theoretical curves for each shape using the intramolecular interference factor $P(\theta)$. Herein, $P(\theta)$ can be expressed by the following Eq. 9.3:

$$P^{-1}(\theta) = \frac{Kc}{R_\theta} \cdot M \tag{9.3}$$

The measured value of $P^{-1}(\theta)$ at each angle is calculated by substituting the inertial radius obtained from the Zimm Plot method into Eq. 9.2. From Eqs. 9.2 and 9.3, Fig. 9.3 is obtained by plotting $P^{-1}(\theta)$ as a function of $\sin^2(\theta/2)$ in both the measured and theoretical curves. Here, the theoretical equations of each shape are summarized in Table 9.1, and the size of the molecular aggregate can be determined by fitting the measured values and the theoretical curve. Furthermore, the most correlated curve with the measured value can be estimated as the shape of the molecular aggregate. In the case of Fig. 9.3, it can be presumed that the shape of this molecular aggregate is ellipsoidal since the measured value has good correlation with the ellipsoidal theoretical curve.

9.4 Understanding Your Data

Equations 9.1 and 9.2 above are general equations for general microparticles. However, when applied to the micelle formed by surfactants, it is necessary to pay attention to the critical micelle concentration (cmc). In this case, Eqs. 9.1 and 9.2 are replaced by Eqs. 9.4 and 9.5, respectively:

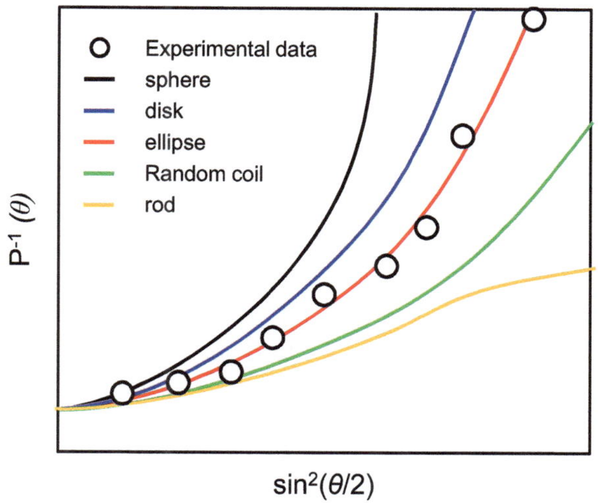

Fig. 9.3 Comparison with experimental data and theoretical curve

$$\frac{K(c-c_0)}{R_\theta - R_0} = \frac{1}{M} + 2A_2(c-c_0) \tag{9.4}$$

$$\frac{K(c-c_0)}{R_\theta - R_0} = \frac{1}{M}\left(1 + \frac{16\pi^2}{3\lambda^2}\left(R_g{}^2\right)\sin^2\left(\frac{\theta}{2}\right) + \ldots\right) + 2A_2(c-c_0) \tag{9.5}$$

C_0: Critical micelle concentration (cmc) [g/ml]
R_0: Inertial radius at cmc [nm]

9.5 Useful Hints

The Debye Plot method can determine the aggregation number and the second virial coefficient only from the angle dependency of the scattered light. However, when the measured particles are larger than 1/20 of the laser wavelength, the Debye Plot method cannot be used.

In the Zimm Plot method, a parallelograms figure is drawn from various measurement angles and concentration as shown in Fig. 9.2. In the following cases, it is necessary to remeasure in order to prevent erroneous analysis; the plot of the measured value protrudes from the parallelogram, and the vertical axis intercept (molecular weight) is negative.

Table 9.1 Theoretical formula in various shape particles

Shape	Shape factor		Inertial radius	Remarks
Sphere	$P(\theta) = \left[\frac{3}{x^3}(\sin x - x\cos x)\right]^2$	$x = \frac{4\pi r}{\lambda}\sin\frac{\theta}{2}$	${R_G}^2 = \frac{3r^2}{5}$	r: radius
Rod	$P(\theta) = \frac{1}{y}\int_0^{2y}\frac{\sin t}{t}dt - \left(\frac{\sin y}{y}\right)^2$	$y = \frac{2\pi l}{\lambda}\sin\frac{\theta}{2}$	${R_G}^2 = \frac{L^2}{12}$	L: length
Random coil	$P(\theta) = \frac{2}{z^2}(z - 1 + e^{-z})$	$z = \frac{16\pi^2}{\lambda^2}\left(\frac{R^2}{6}\right)\sin^2\frac{\theta}{2}$	${R_G}^2 = \frac{R^2}{6}$	R: average distance between both ends
Disk	$P(\theta) = \frac{2}{X^2}\left\{1 - \frac{J_1(2X)}{X}\right\}$	$X = \frac{4\pi r}{\lambda}\sin\frac{\theta}{2}$	${R_G}^2 = \frac{r^2}{2}$	r: radius
Ellipse	$P(\theta) = \frac{9\pi}{2}\int_0^{\frac{\pi}{2}}\frac{{J_{3/2}}^2(V)}{V^3}\cos\beta\ d\beta$	$V^2 = \left(\frac{4\pi}{\lambda}\sin\frac{\theta}{2}\right)^2 \cdot (a^2\cos^2\beta + b^2\sin^2\beta)$ $J_{3/2}(V) = \frac{2a^2 + b^2}{5}$	${R_G}^2 = \frac{2a^2+b^2}{5}$	a: short axis length, b: long axis length

Chapter 10
Dynamic Light Scattering (DLS)

Takeshi Misono

Abstract Particles dispersed in a solution are constantly moving in Brownian motion. Since this Brownian motion depends on the velocity of the particle size, the size can be estimated by observing the fluctuation corresponding to this velocity. By obtaining the autocorrelation function of the observed fluctuation, the particle size and distribution can be obtained. Dynamic light scattering (DLS) uses the above principle and provides size information of a wide range of particles easily and accurately. Since the size information of particles, molecular aggregates, and emulsion droplet is extremely important, DLS is widely preferred by many colloid researchers. However, data analysis of DLS is based on the assumption that the particles are uniform and spherical. It should also be noted that the obtained size includes a solvated layer around the particles. Due to its nature, other equipment must be used to obtain detailed information such as the shape. The principle and detailed analysis method by DLS measurement will be described in this section.

Keywords Dynamic light scattering · Particle size · Hydrodynamic radius · Brownian motion · Stokes-Einstein equation

10.1 Introduction

Particles (micelles, microparticles, etc.) in colloidal dimension size (c.a. 1 μm to 1 nm) are constantly moving in solutions due to translational diffusion in Brownian motion. When light is irradiated, the fluctuation of the scattering intensity that depends on the particle size in the solution can be observed. In other words, small particles diffuse quickly and the scattering intensities fuctuate quickly, while large particles fluctuate slowly. By analyzing the autocorrelation based on this difference variation, the diffusion coefficient can be obtained, and from this the particle size can be obtained. Therefore, dynamic light scattering (DLS) is a measurement that can

T. Misono (✉)
NIKKOL Group Cosmos Technical Center Co., Ltd., Tokyo, Japan
e-mail: t.misono@ns-cosmos.co.jp

M. Abe (ed.), *Measurement Techniques and Practices of Colloid and Interface Phenomena*, https://doi.org/10.1007/978-981-13-5931-6_10

easily obtain the size information for a molecular aggregate formed by surfactants (typically micelles), microparticles, etc.

10.2 What You Get

1. Measurement of a wide range of particle sizes from 1 nm to 5 μm
2. Noncontact measurement using light, allowing undamaged sample measurement
3. Particle size, if the viscosity and refractive index of solvent are known

10.3 Essentials and Tips

Figure 10.1 shows the principle of dynamic light scattering measurement and a simple process to determine the particle size. As mentioned in the introduction section, the particles are constantly diffusing in the solution in Brownian motion. Therefore, time-dependent spatial position change of the particles causes the variation of the light scattering intensity due to interference. By analyzing autocorrelation from the temporal fluctuation of the scattering intensity, the diffusion coefficient is obtained. Assuming that the molecules are spherical and uniform, the hydrodynamic radius can be determined by the Stokes-Einstein equation (Eq. 10.1):

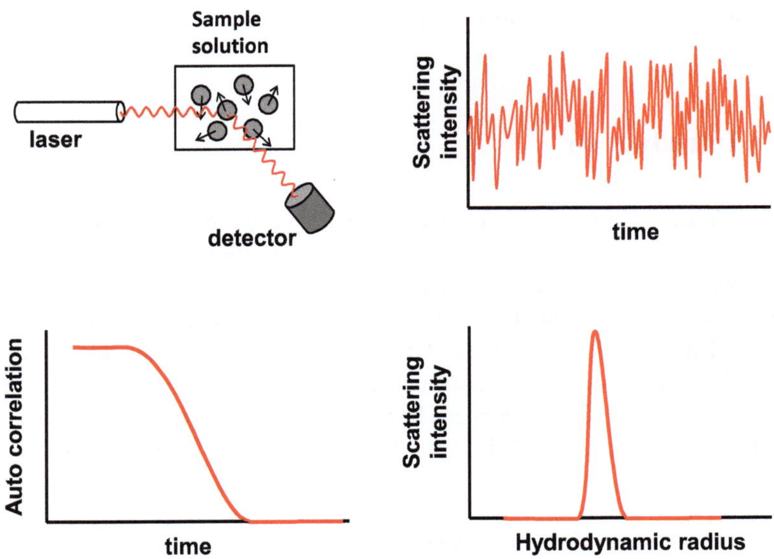

Fig. 10.1 Process from dynamic light scattering measurement to obtaining particle size distribution

$$D = \frac{K_B T}{3\pi\eta d} \tag{10.1}$$

D: Diffusion coefficient
K_B: Boltzmann constant
T: Temperature
η: Viscosity
d: Hydrodynamic diameter

From the particle size information, the formation of molecular aggregate can be confirmed.

10.4 Understanding Your Data

10.4.1 Quality of Autocorrelation Function Graph

It is common to evaluate the particle size from the calculated size distribution graph (right bottom in Fig. 10.1); the autocorrelation function graph must be evaluated carefully. Therefore, it is important to judge whether the autocorrelation function graph is eligible for the analysis. Figure 10.2 shows examples of autocorrelation function graphs, and their interpretation are explained as follows:

The autocorrelation curve in Fig. 10.2a has decays with two inflection points, indicating the existence of two different size particles in the sample with fast attenuation due to small particles and slow attenuation due to large particles.

The autocorrelation curve in Fig. 10.2b stays unchanged for a while and then attenuates, though it would not completely decay even after a long time period. This curve indicates the influence of scattering by considerably large particles due to contamination of bubbles and dust. If such a curve is seen, re-measuring is required after a centrifugation or filtering process.

An irregular function is observed early in the attenuation curve in Fig. 10.2c. This phenomenon is often observed when the sole solvent does not have particles. Therefore, curve type in Fig. 10.2c indicates that the particle sizes are too small or the particle concentration is too dilute. The laser intensity must be increased to increase the scattered light intensity from the small particles or to increase particle concentration.

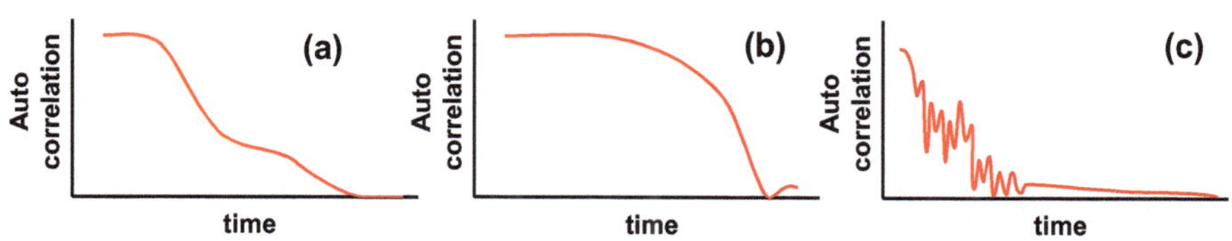

Fig. 10.2 Autocorrelation function graph requiring attention

10.4.2 Types of Particle Size Distribution

There are three types of size distribution obtained from DLS: scatter intensity, volume, and number distribution. Figure 10.3 shows the difference between volume and number distribution. The scattered light intensity depends on the particle volume. The large particles give strong scattered intensity even in small amounts because the radius cubed equals the volume. Hence, it is important to understand what the objective information required from DLS measurement is, and volume distribution is often used for molecular aggregates by surfactants, while number distribution is used for bulky monomers and solid microparticles.

10.5 What to Look Out for

1. Data analysis of DLS is based on the assumption that the particles are uniform and spherical. Because of its feature, other equipment is required to obtain detailed information such as the shape. Additionally, it is important to keep in mind that the obtained size includes a solvated layer around the particles.
2. The viscosity and refractive index of the solution are required for accurate measurement, since Stokes-Einstein equation (Eq. 10.1) is used for analysis.
3. When foreign substances (dust, bubbles, etc.) enter the measurement sample, a stronger scattering intensity than the target particle is detected. Therefore, it is important to prevent contamination of foreign substances by using a filter.
4. When the sample solution is cloudy at high concentration, it is affected by multiple scattered lights. Therefore, it is preferable that the sample concentration is as low as possible.

10.6 Useful Hints

Finally, NICOMP will be described among a number of dynamic light scattering equipment in here. The instrument is depicted in Fig. 10.4. NICOMP has flexible customizability and is also equipped with automatic sample changer. Additionally,

r	1	5	10	20
r^3	1	125	1000	8000
n	1000	200	100	5
n (%)	76.6	15.3	7.7	0.4
V (%)	0.6	15.1	60.2	24.1

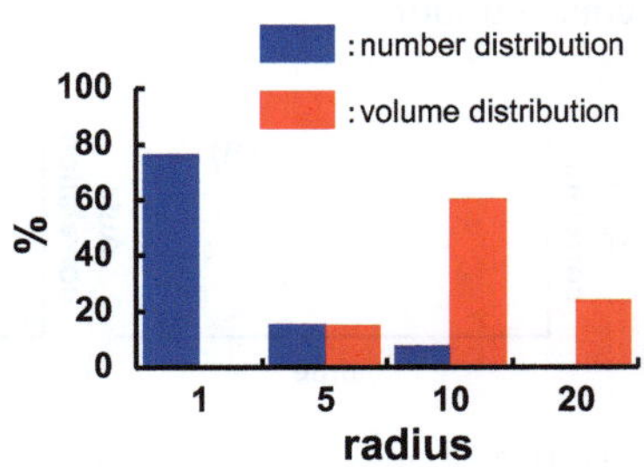

Fig. 10.3 Difference between number and volume distribution

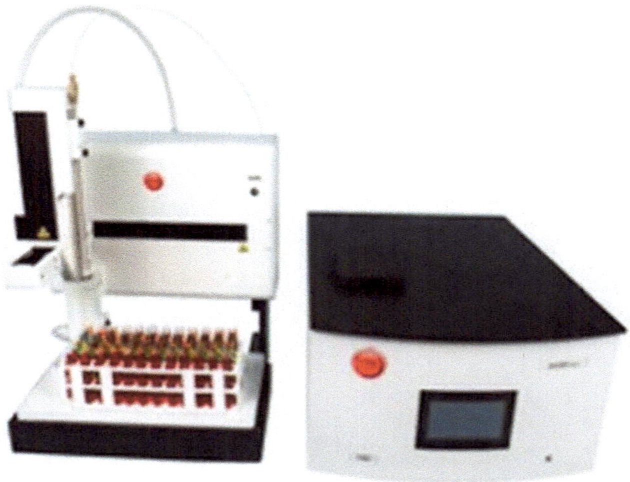

Fig. 10.4 Dynamic light scattering equipment (NICOMP N3000) and autosampler. (Photo credit Nihon Entegris K.K.)

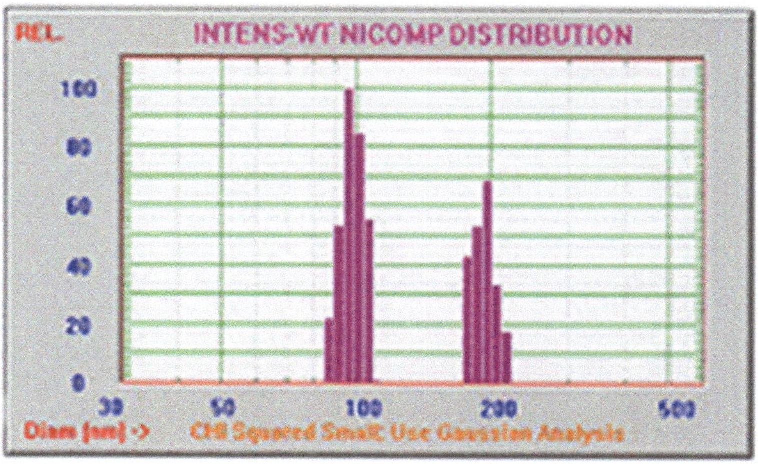

Fig. 10.5 Mixture of two standard latex particles with 93 and 150 nm

its unique analysis software "NICOMP algorithm" is included for particle size analysis, so accurate measurement samples with multiple peaks can be performed. Figure 10.5 shows the results of mixed samples with different size particles, i.e., 93 and 150 nm, using NICOMP. Despite the relatively small size differences, NICOMP can provide accurate particle size and distribution.

Chapter 11
Solubilization by Micelles

Masahiko Abe

Abstract Solubilization resembles emulsification in regard to how they are caused by surfactant self-assemblies. Here, the distinct difference is that solubilization is a state under thermodynamic equilibrium and thus is independent to the process applied, but on the other hand, emulsion is an unstable non-equilibrium state highly dependent on the preparation process. Systems called microemulsions require special attention because despite of the name, they are states in solubilization where swelled micelles are dispersed. The term solubilization is generally used for isotropic single-phase micelle solutions higher than the Kraft temperature, though there are recent trends to include lyotropic liquid crystals and gels.

The maximum solubilization by micelles for slightly soluble substances at certain temperature is quite important to measure. Therefore, this chapter deals with solubilization of volatile and/or nonvolatile substances by surfactant micellar solution.

Keywords Solubilization · Thermodynamic equilibrium · Micelle · Micro emulsion · Emulsification

11.1 Introduction

Solubilization by micelles can generally be categorized as a phenomenon called sorption, and further divided as adsorption onto the micelle surface and absorption into the micelle core. Solubilization is increasing the solubility of relatively immiscible materials in solvents, in most cases water, by adding surfactant(s) to the system. Solubilization is a thermodynamically stable phenomenon of making single-phase solutions regardless of the process applied and one of the important properties of surfactants such as emulsification or dispersibility. The term solubilization is usually used for isotropic single-phase micelle solutions higher than the Kraft temperature, though there are recent trends to include lyotropic liquid crystals and gels. Therefore,

M. Abe (✉)
Research Institute for Science and Technology, Tokyo University of Science, Noda, Japan
e-mail: abemasa@rs.noda.tus.ac.jp

M. Abe (ed.), *Measurement Techniques and Practices of Colloid and Interface Phenomena*, https://doi.org/10.1007/978-981-13-5931-6_11

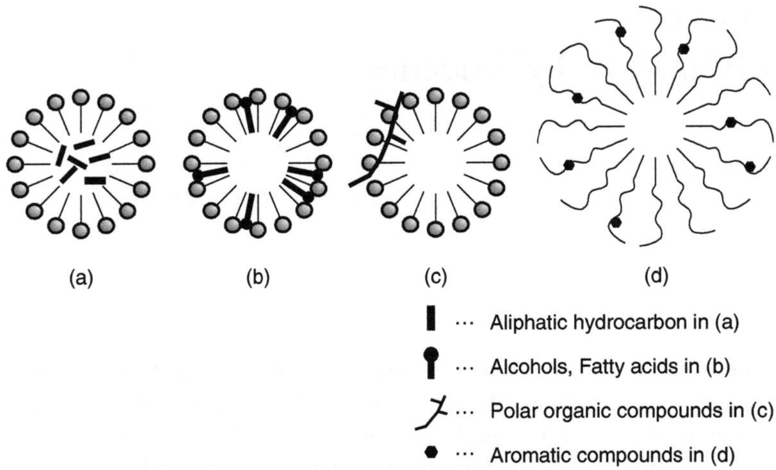

Fig. 11.1 Solubilization position of various compounds in surfactant micelles

solubilization could be regarded as a property of surfactant solutions to increase the solubility of organic materials. There is a similar phenomenon called hydrotropy, which can be observed when adding alcohol or acetone to water. This is distinguished from solubilization due to the lack of micelles or self-assemblies. Solubilization also resembles emulsification in regard that both self-assemblies are caused by surfactant. Here, distinct differences are that solubilization is a state under a thermodynamic equilibrium and is independent to the process applied, whereas emulsion is an unstable non-equilibrium state highly dependent on the preparation process. Systems called microemulsions require special attention because despite of the name, they are states in solubilization where swelled micelle is dispersed.

There are four types of solubilizations by micelles (a)–(d) as shown in Fig. 11.1, based on the position of solubilizate in the micelle. For (a), hydrophobic materials such as aliphatic or aromatic hydrocarbons are solubilized in the micelle core at the hydrocarbon tails of surfactant, (b) is cases when fatty alcohol, fatty acids, and fatty amines are solubilized in between the surfactant hydrocarbons nearby the micelle surface called palisade, (c) is when polar organic materials and dyes with rather large molecules are solubilized at the micelle surface, and (d) is cases where water soluble materials with affinity to surfactant hydrophilic moiety are solubilized at the surface hydrosphere of surfactant. The approximate solubilization capacity is generally recognized as (d) > (b) > (a) > (c). Micelles swell when materials are solubilized and are more noticeable for substances with polar groups. Solubilization by reversed micelles occurs in polar materials and water.

11.2 Solubilization for Nonvolatile Materials

The amount of solubilized nonvolatile dyes can be measured by optical absorption at the maximum wavelength of the dye molecule. Solubilization of molecules without absorption can be measured by the opacity of the solution at around 700 nm. The actual method is measuring the absorption of the varied dye or material concentration solubilized in micelles with a fixed surfactant concentration. The determined absorption values are plotted against the dye or material concentration, which linearly increases with concentration and then deviates from the line as shown in Fig. 11.2. The concentration when the deviation starts is the maximum solubilization (Smax) of the micelle at the given surfactant concentration. Next, Smax are plotted to the surfactant concentration as shown in Fig. 11.3. The solubilization capacity can be obtained from the slope of the line, and the intercept point corresponds to the cmc of the surfactant.

11.3 What to Look Out for

The Smax of solubilizates will be different depending on the method applied, i.e., either batch or continuous. The batch method consists of multiple surfactant solutions with same concentration. Dyes are added to each surfactant solution in incremental amounts to each solution in the beakers. In the continuous method,

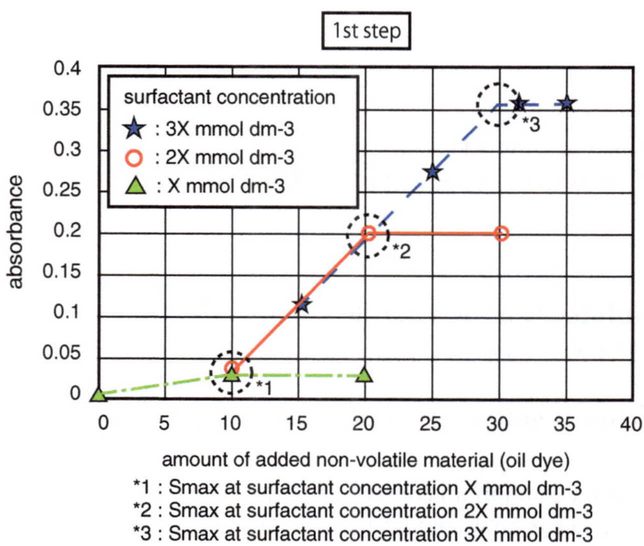

Fig. 11.2 Determination of the maximum solubilization (Smax) at the fixed surfactant concentration (conceptual drawing)

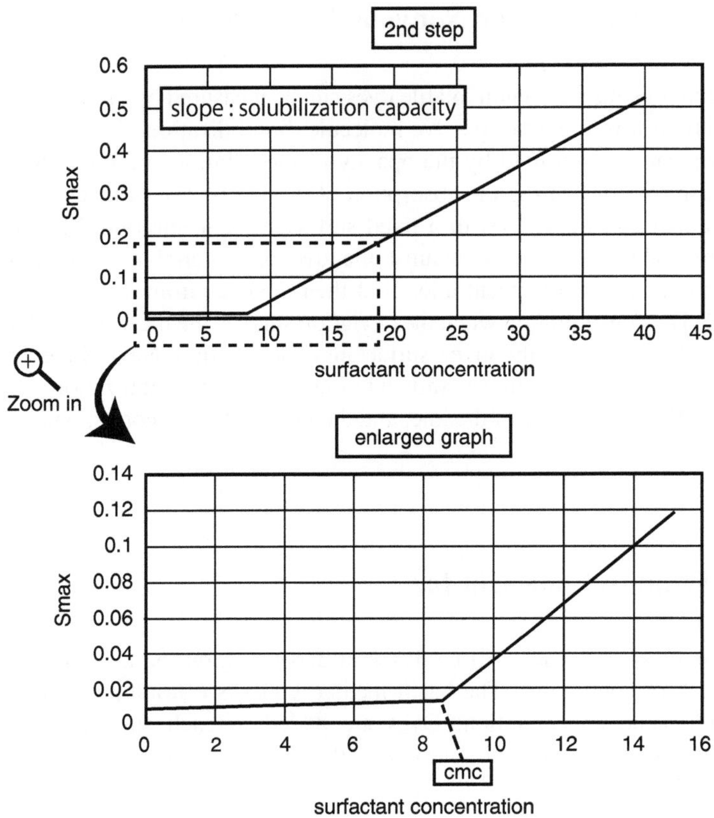

Fig. 11.3 Maximum solubilization (Smax) by the micelle at given surfactant concentration (conceptual drawing)

small amounts of dye are added to solubilize successively like titration. As a result, the Smax of the continuous method is larger than the batch method.

11.4 Solubilization of Volatile Chemicals Such as Fragrances

11.4.1 Determination of Solubilization Equilibrium Constant

The solubilization equilibrium constant (*K*: L/mol) represents the partition constant of volatile materials in between micelles and the bulk at the equilibrium. *K* is a ratio of the mole fraction of the solubilized volatile in the micelle (X_{org}) and in the bulk (C_{org}) as shown in Eq. 11.1.

$$K = \frac{X_{\mathrm{org}}}{C_{\mathrm{org}}} \tag{11.1}$$

A larger K value indicates more solubilization of volatile in the micelle, and the dimension of K is (M^{-1}). Generally, the Hansen-Millar equation [1] expresses a linear relationship between the activity coefficient, in this case the solubilization constant K and molarity X_{org} [2]. This is true for the volatile molecule when X_{org} is lower than 0.3. However, when X_{org} is higher than 0.3–0.6, a better relationship is known as shown in Eq. 11.2 [3–6]:

$$\sqrt{K} = \sqrt{K_0} - B\sqrt{K_0}X \tag{11.2}$$

where K_0 is extrapolated K at $X = 0$ and B is a constant value. The activity coefficient of volatile in the micelle is shown as Eq. 11.3. This empirical relationship is well accepted for many volatile chemicals solubilized in micelles [3–5].

$$\gamma_{\mathrm{org}} = \frac{1}{KC_{\mathrm{org}}} = \frac{A}{\left(1 - BX_{\mathrm{org}}\right)} \tag{11.3}$$

where $A = \frac{1}{K_0 C^0_{\mathrm{org}}}$

11.5 Static Headspace Method [7, 8]

The static headspace method, or headspace gas chromatography, is a method to evaluate the solubilization of volatiles in solutions by measuring the partial vapor pressure of the volatile in the gas phase. Under an equilibrium condition of gas and liquid phase, the partial vapor pressure of the target molecule will increase with increasing the concentration of the molecule in the liquid so the partial vapor pressure corresponds to the concentration in the solution.

There are two types of the headspace method, dynamic and static. With the dynamic headspace method, gases with a target volatile are evaporated through a looped apparatus and then absorbed onto the absorption resin or bed for a fixed time period. Then the absorbed molecules are released by various methods and are quantitatively analyzed by gas chromatography. Kinetic measurement of the rate of evaporation is possible with this dynamic method.

The static method consists of the following processes. First, aqueous solutions or solids with volatile materials are put into sealed container to hold still until they reach the equilibrium under a constant temperature, and then the gas at the headspace of the container is analyzed by gas chromatography. This static headspace method is commonly applied to the analysis of volatile materials in foods and fragrances. To apply this method, solubilized oily materials in the specimen must be volatile, and

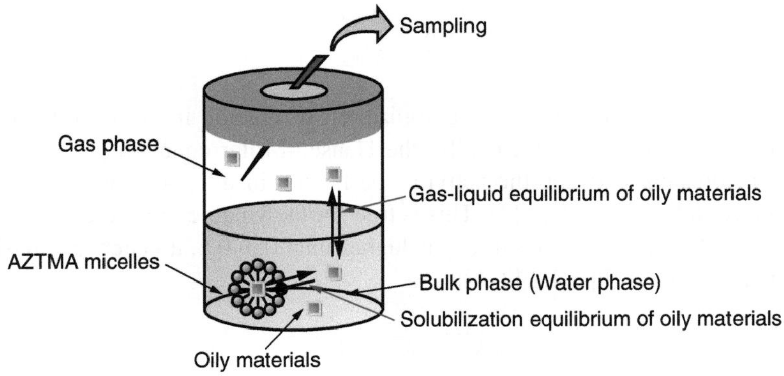

Fig. 11.4 Schematization of solubilizing solution

two continuous partition equilibriums exist in the system as shown in Fig. 11.4. One is a solubilization equilibrium of volatile molecules between the micelle and bulk solution, and the other is a gas (headspace) and liquid partition equilibrium of the molecule.

The partial vapor pressure of volatiles (solubilizates) in the headspace gas is decreased when the solubilized amount of volatile in the micelle is increased as these two equilibriums are interrelated to each other. The partial vapor pressure of the volatile molecule is equivalent to the molar amount of the molecule in the headspace. Therefore, the solubilization amount at each volatile concentration in the solution or the maximum solubilization (Smax) can be indirectly measured by determining the partial vapor pressure of the volatile in the headspace.

11.6 Experiment Procedure for the Static Headspace Method

Sealed vials with solutions of fixed surfactant concentration and varied amounts of volatiles (solubilizates) are set to reach equilibrium. The headspace gases of each vial are sampled by gas-tight syringes and analyzed by gas chromatography. This process when below or above the maximum solubilization (Smax) is shown in Fig. 11.5, and the relationship between the partial vapor pressure (P/P_0) and Smax is shown in Fig. 11.6. In order to correlate the solubilization equilibrium of volatiles between micelles and bulk solutions to the gas-liquid equilibrium, the activity coefficient of volatiles in the aqueous solutions must be determined. Additionally, the solubilization capacity can be determined from the slope of the line by plotting the Smax at different surfactant concentrations against the surfactant concentration.

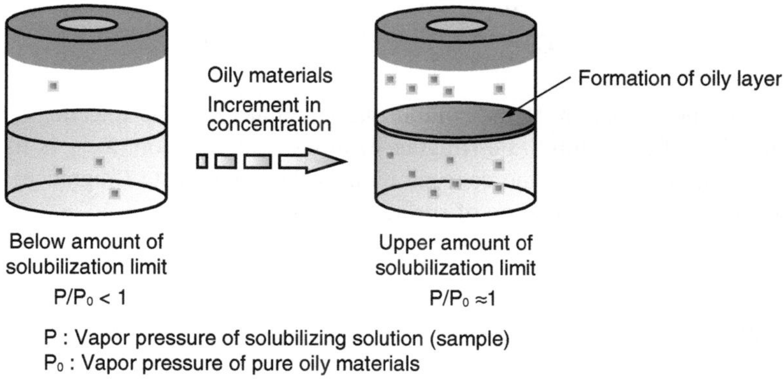

Fig. 11.5 Relationship between vapor pressure and solubilizing amount

Fig. 11.6 Relationship between vapor pressure and concentration of volatiles

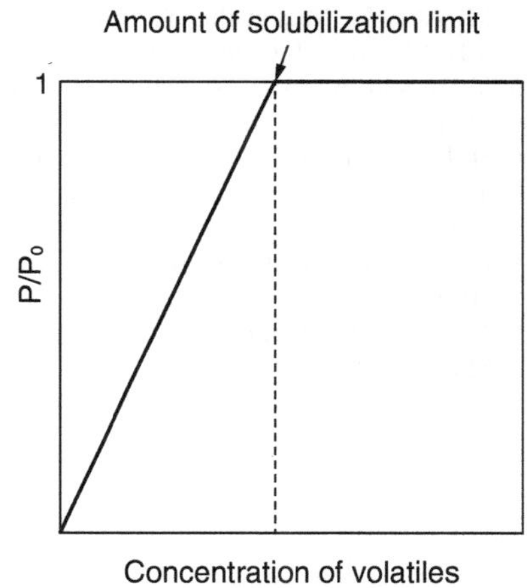

11.7 An Alternative Method to Determine the Solubilization Equilibrium Constant (*K*)

The following is a short note of the semi-equilibrium dialysis (SED) method for determining *K* [3, 4, 9], in which details can be found in the original papers referred. Solubilized solutions are placed in the retentate of the dialysis cell while pure water is set into the permeate, which are separated by a cellulose membrane with pores to cut molecules over 6000 Dalton off the dialysis. After 24 h at 30 °C, the concentrations of the surfactants and volatile solubilizates in the permeate part of the cell are analyzed. *K* can be determined with Eq. 11.4

$$K = \frac{X}{C} = \frac{X}{55.X_{\text{bulk}}} \quad \text{and} \quad X = (Cm)/(Csf + Cm) \qquad (11.4)$$

where X is mole fraction of the solubilizate (fragrance) in the micelle, C is the concentration of the solubilizate (fragrance) in the bulk solution, X_{bulk} is its mole fraction in the bulk, Cm is the concentration of the solubilizate (fragrance) in the micelle, and Csf is concentration of surfactant in solution.

References

1. R.S. Hansen, F.A. Millar, J. Phys. Chem. **58**, 193 (1954)
2. G.A. Smith, S.D. Christian, E.E. Tucker, J.F. Scamehorn, J. Solut. Chem. **15**, 519 (1986)
3. B.H. Lee, S.D. Christian, E.E. Tucker, J.F. Scamehorn, Langmuir **6**, 230 (1990)
4. M. Abe, K. Mizuguchi, Y. Kondo, K. Ogino, H. Uchiyama, J.F. Scamehorn, E.E. Tucker, S.D. Christian, J. Colloid Interface Sci. **160**, 16 (1993)
5. Y. Kondo, M. Abe, K. Ogino, H. Uchiyama, J.F. Scamehorn, E.E. Tucker, S.D. Christian, Langmuir **9**, 899 (1993)
6. H. Uchiyama, S.D. Christian, J.F. Scamehorn, M. Abe, K. Ogino, Langmuir **7**, 95 (1991)
7. T. Shikata, Y. Sakaiguchi, H. Urakami, A. Tamura, H. Hirata, J. Colloid Interface Sci. **133**, 288 (1989)
8. M.E. Morgan, H. Uchiyama, S.D. Christian, E.E. Tucker, J.F. Scamehorn, Langmuir **10**, 2170 (1992)
9. Y. Kondo, K. Mizuguchi, Y. Tokuoka, H. Uchiyama, K. Kamogawa, M. Abe, J. Jpn Soc. Colour Mater **68**, 271 (1995)

Chapter 12
Rheology

Koji Tsuchiya

Abstract Rheology is the study of flow behavior and the subsequent deformation of materials as a result of flow. Viscosity and elastic modulus of materials are measured in rheology. The science of rheology and the characterization of viscoelastic behavior are widely used for producing many industrial products including detergents, shampoos, toothpastes, cosmetics, paints, and foods. The measurement of the flow curve is important for the first step of rheological measurements. In flow curves, the shear stress of a sample is plotted against the shear rate. Newtonian fluids are characterized by a single viscosity where the shear stress is proportional to the shear rate. Only a few of materials show this behavior. Most materials behave as non-Newtonian fluids, in which viscosity changes with the shear rate. In some cases, the rheological parameters measured with a rheometer or a viscometer do not match with the actual phenomena. This is mainly because measurement conditions are inappropriate for evaluating the actual phenomena. In rheology, it is necessary to appropriately select measurement method, conditions, and apparatuses.

Keywords Rheology · Flow curves · Dynamic viscoelasticity · Thixotropy · Cole-Cole plots

12.1 Introduction

Rheology is a science to evaluate phenomena related to stress, strain (deformation), and time. There are many rheological phenomena involved in familiar things such as painting [1, 2], feel of cosmetics [3–5], and texture of food [5, 6].

The viscosity and elastic modulus of materials are measured in rheology. However, it is often difficult to understand what these parameters mean. In some cases, the results measured with a rheometer or a viscometer do not match with the actual phenomena. For example, when painting on a wall, there are cases where paint drips

K. Tsuchiya (✉)
Research Institute for Science and Technology, Tokyo University of Science, Noda, Japan
e-mail: kjtsuchi@rs.noda.tus.ac.jp

M. Abe (ed.), *Measurement Techniques and Practices of Colloid and Interface Phenomena*, https://doi.org/10.1007/978-981-13-5931-6_12

even though the viscosity value is high. This is mainly due to inappropriate measurement conditions for evaluating the actual phenomena. In rheology, it is necessary to appropriately select the method of measurement, the conditions, and the measurement apparatus according to the sample and phenomenon.

This section focuses mainly on methods and important points on rheological measurements using elementary mathematical formulas and principles. For details on mathematical formulas and analytical methods, refer to the specialized textbooks.

12.2 What You Get

1. Information on rheological parameters such as the viscosity and elastic modulus.
2. Information on the thixotropy of a sample. Thixotropy [7] is a property where viscosity changes with time. Thixotropy is seen particularly in dispersion systems. When the stirring rate is high, the viscosity of the thixotropic sample decreases into a liquid state. In contrast, when the stirring rate is low, the viscosity increases, and the sample behaves like a solid state. Application of thixotropy includes paints and toothpaste.

12.3 Steps and Essentials

Elastic modulus and viscosity are the most fundamental parameters in rheological measurements. The elastic modulus is based on Hooke's law, where stress is proportional to strain. The elastic modulus is expressed by

$$E = \frac{\sigma}{\varepsilon}$$

where E is elastic modulus (Young's modulus), σ is stress, and ε is strain.

When weak stress is applied, the material deforms (strains) due to the stress and returns to the original state if the stress is removed. This region is called the linear region. When applying more stress, the material does not return to its original state with plastic deformation even if the stress is removed. Viscoelasticity measurements must be carried out within the linear region. In some cases, the plastic deformation might occur from the beginning of stress, and the elastic modulus might not be clearly defined. When a viscous fluid flows, a displacement occurs between the contacting layers, and a tangential stress is generated during this shift. The shear stress (σ) is proportional to the shear strain per unit time, that is, the shear rate ($\dot{\gamma}$). The proportional coefficient of the σ–$\dot{\gamma}$ curve is the viscosity (η).

$$\eta = \frac{\sigma}{\dot{\gamma}}$$

Materials that follow this equation are called Newtonian fluids. In contrast, materials whose viscosity depends on shear rate are called non-Newtonian fluids.

12.3.1 Selection of Viscosity-Measuring Apparatus

When the measurement temperature is around room temperature, E-type viscometers or rheometers equipped with a cone-plate geometry are used. The amount of the sample should be sufficient and fill the gap between the cone and plate. If the sample protrudes from the diameter of the cone, it is overfilled so the excess sample should be cleanly removed with a spatula or filter paper. Reproducibility cannot be obtained with too little or too much amount of sample. Since bubbles in the sample cause measurement error, they must be removed before measurement.

If the viscosity of the sample is considerably low, it is difficult to measure viscoelasticity with cone-plate-type rheometers. In this case, rheometers equipped with a double concentric cylinder are used. In contrast, for high viscosity samples, rheometers equipped with a parallel plate are used.

12.3.2 Search and Importance of Linear Region

Dynamic viscoelasticity must be measured within the linear region where the stress is proportional to the strain. In the linear region, the storage modulus G', an index of elasticity, and the loss modulus G'', an index of viscosity, are constant independent of stress and strain. In the nonlinear region, the result is not reproducible, and it is difficult to understand the inherent physical properties of the sample. When measuring with a rheometer, attention must be paid since the rheological parameters are obtained even in nonlinear regions. For unknown samples, the linear region must be determined before dynamic viscoelasticity measurement.

Lissajous figures [8] are useful for determining the linear region. The horizontal axis represents the strain periodically varied, and the vertical axis plots the response stress to the strain. Figure 12.1 shows Lissajous figures of a purely elastic material, a purely viscous material, and a viscoelastic material. Since the purely elastic material follows the Hooke's law, the Lissajous figure becomes a straight line at an angle of 45°. For the purely viscous material, the Lissajous figure draws a circle. Since a viscoelastic material behaves in the middle between elastic and viscous materials, its Lissajour figure shows an ellipse shape tilted at an angle of 45°. If the Lissajour figure is not such an ellipse shape, the measurement condition lies in the nonlinear region.

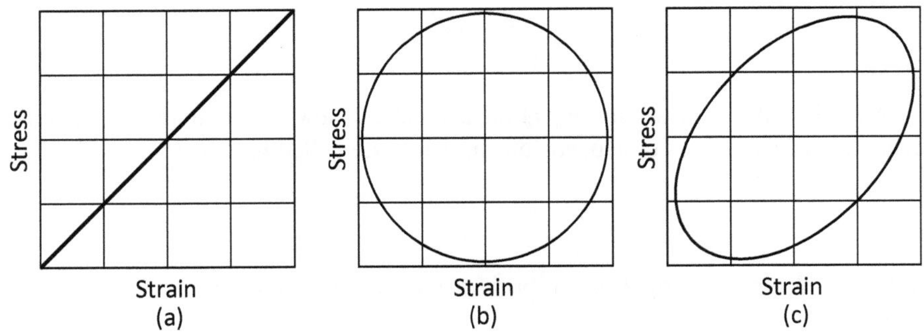

Fig. 12.1 Lissajous figures of (**a**) a purely elastic material, (**b**) purely viscous material, and (**c**) viscoelastic material

Fig. 12.2 Linear and nonlinear regions

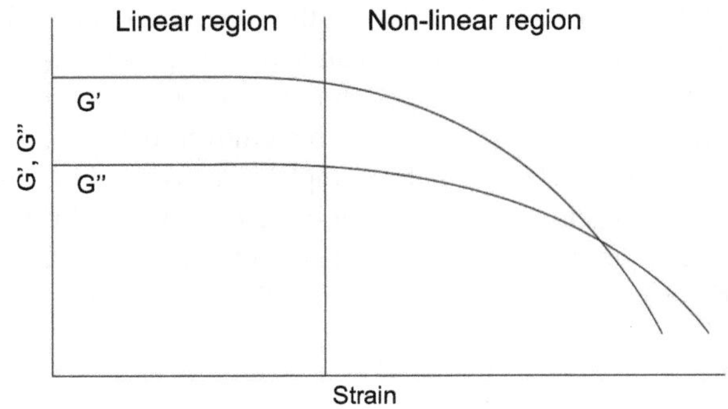

As another method for conveniently confirming the linear region, the strain (or stress) dependence of G' and G'' is measured at a fixed frequency, usually 1 Hz (Fig. 12.2). The linear region is a region in which G' and G'' are constant.

12.4 Understanding Your Data

12.4.1 Flow Curves

The measurement of a flow curve is a static measurement, where shear strain is applied to the sample fluid in one direction, and it is important for the first step of rheological measurements. In the flow curve, the vertical axis and horizontal axis represent shear stress and shear rate, respectively. Usually the vertical and horizontal axes are shown logarithmically.

Figure 12.3 shows a schematic representation of various types of flow curves. The flow curve is classified into the following according to its shape:

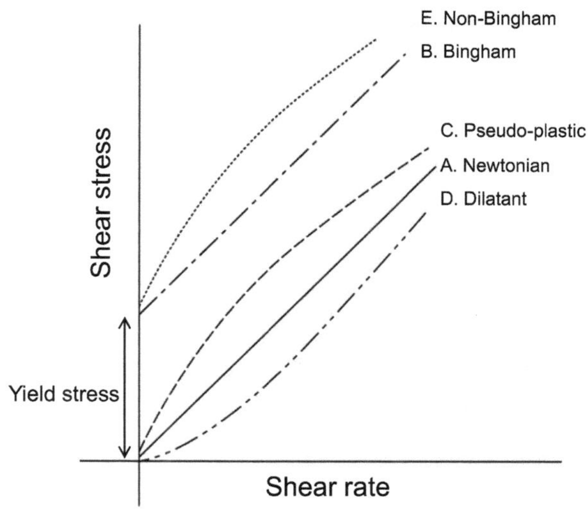

Fig. 12.3 Various types of flow curves

A. Straight line passing through the origin (Newtonian behavior)
B. Straight line not passing through the origin and having the yield stress (Bingham behavior)
C. Convex-upward curve passing through the origin (pseudoplastic behavior)
D. Convex-downward curve (dilatant behavior)
E. Convex-upward curve not passing through the origin and having the yield stress (non-Bingham behavior)

12.4.2 Dynamic Viscoelasticity

In dynamic viscoelasticity measurements, the viscoelasticity of a sample is measured by applying periodically changed strain or stress. The storage modulus G' and loss modulus G'' are indices of elasticity and viscosity, respectively. In a typical frequency sweep measurement, the applied strain (or stress) is controlled to be constant within the linear region, and the G' and G'' are obtained.

The phase difference of the stress σ responsive to the sine-waved strain $\gamma = \gamma_0 \sin \omega t$ is represented by δ. In the case of a purely elastic material, since there is no phase difference between strain and stress, δ becomes zero. On the other hand, the phase difference between strain and stress is shifted by $\pi/2$ for a purely viscous material. In the case of a viscoelastic material, δ shows a value between 0 and $\pi/2$.

Pure elastic material; $\sigma = \sigma_0 \sin \omega t$
Pure viscous material; $\sigma = \sigma_0 \sin (\omega t + \pi/2)$
Viscoelastic material; $\sigma = \sigma_0 \sin (\omega t + \delta)$ $(0 < \delta < \pi/2)$

The complex modulus of elasticity E^* is stress divided by strain and is shown as follows:

$$E^* = G' + iG''$$
$$G' = \frac{\sigma_0}{\gamma_0}\cos\delta$$
$$G'' = \frac{\sigma_0}{\gamma_0}\sin\delta$$

γ_0 and σ_0 are the amplitude of strain and the amplitude of stress, respectively.

The loss tangent (tan δ) indicates the ratio of the viscosity term based on the elastic term.

$$\tan\delta = \frac{G''}{G'}$$

When tan δ is less than 1, G' becomes larger than G'' (elasticity-dominated). In contrast, when tan δ is larger than 1, viscosity becomes dominant ($G' < G''$).

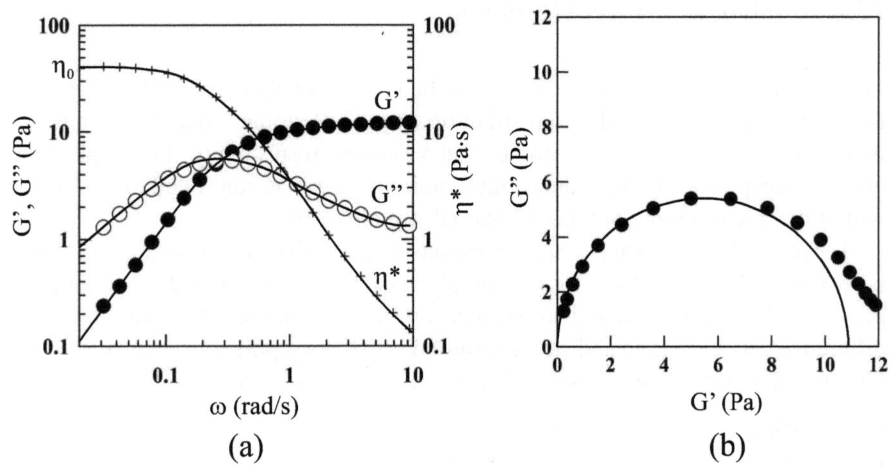

Fig. 12.4 (**a**) Storage modulus G', loss modulus G'', and complex viscosity η^* as a function of angular frequency ω and (**b**) Cole-Cole plots for a worm-like micellar system. The semicircle draws the behavior of a Maxwell model

12.5 Useful Hints

Worm-like micellar systems exhibit a remarkable viscoelastic behavior due to the formation of a three-dimensional network consisting of entangled micelles. Figure 12.4 shows the storage modulus G', the loss modulus G'', and the complex viscosity η^* as a function of angular frequency ω. The G' curve exhibits a definite plateau at high frequencies, and G'' is higher than G' at low frequencies. This viscoelastic behavior is consistent with a Maxwell model, in which a purely viscous material and a purely elastic material are connected in series. It is well-known that worm-like micellar solutions exhibit viscoelastic behavior similar to a Maxwell model [9–14]. The zero-shear viscosity η_0, which is given by $\lim_{\omega \to 0} \eta^*$, can be obtained from the η^* curve. Figure 12.4b shows the relationship between G'' and G', a Cole-Cole plot. For a Maxwell model, its Cole-Cole plot shows a semicircle-shaped curve.

References

1. P. Snabre, P. Mills, Colloids Surf. A **152**, 79–88 (1999)
2. G.E. Morris, W.A. Skinner, P.G. Self, R.S.C. Smart, Colloids Surf. A Physicochem. Eng. Asp. **155**, 27–41 (1999)
3. C. Valenta, K. Schultz, J. Control. Release **95**, 257–265 (2004)
4. D. Bais, A. Trevisan, R. Lapasin, P. Partal, C. Gallegos, J. Colloid Interface Sci. **290**, 546–556 (2005)
5. C. Gallegos, J.M. Franco, Curr. Opin. Colloid Interface Sci. **4**, 288–293 (1999)
6. H.M. Shewan, J.R. Stokes, J. Food Eng. **119**, 781–792 (2013)
7. F. Bautista, J.M. De Santos, J.E. Puig, O. Manero, J. Non-Newtonian Fluid Mech. **80**, 93–113 (1999)
8. H.A.H. Al-Khazali, M.R. Askari, IOSR J. Eng. **2**, 971–978 (2012)
9. T. Shikata, H. Hirata, T. Kotaka, Langmuir **3**, 1081–1086 (1987)
10. H. Rehage, H. Hoffmann, Mol. Phys. **74**, 933–973 (1991)
11. L.M. Walker, Curr. Opin. Colloid Interface Sci. **6**, 451–456 (2001)
12. K. Tsuchiya, Y. Orihara, Y. Kondo, N. Yoshino, T. Ohkubo, H. Sakai, M. Abe, J. Am. Chem. Soc. **126**, 12282–12283 (2004)
13. H. Sakai, Y. Orihara, H. Kodashima, A. Matsumura, T. Ohkubo, K. Tsuchiya, M. Abe, J. Am. Chem. Soc. **127**, 13454–13455 (2005)
14. H. Sakai, S. Taki, K. Tsuchiya, A. Matsumura, K. Sakai, M. Abe, Chem. Lett. **41**, 247–248 (2012)

Chapter 13
Freeze-Fracture Transmission Electron Microscopy

Koji Tsuchiya

Abstract Freeze-fracture transmission electron microscopy (FF-TEM) gives us useful information about the structure of molecular aggregates, colloidal dispersions, and emulsions. Rapidly frozen samples are fractured with a glass knife. Platinum (Pt) is evaporated and deposited on the fractured surface at 45° to make a thin Pt film. This "shadowing" process induces an enhancement of the contrast of the TEM image depending on the roughness of the fractured surface. Carbon (C) is then evaporated and deposited on the fractured surface at 90° to make a "replica" film. The replica film thus prepared is observed with a transmission electron microscope at high resolution. FF-TEM allows observation of almost all molecular aggregates formed by amphiphilic molecules (surfactants) and colloidal particles like emulsion dispersed in both water and organic solvents. In liposome (vesicle) systems, characteristic TEM images can be obtained depending on the conformation of hydrophobic tails of lipid molecules, that is, gel phase (L_β), ripple gel phase (P_β), and liquid crystal phase (L_α).

Keywords Freeze-fracture TEM · Freeze-etching TEM · Replica films · Rapid freezing

13.1 Introduction

Surfactants form various kinds of molecular aggregates such as spherical micelles, rod-like micelles, vesicles, hexagonal liquid crystals, cubic liquid crystals, and lamellar liquid crystals in water or organic solvents. Since most interfacial phenomena are deeply related to the formation of molecular aggregates of surfactants, it is important to determine the size and shape of the molecular aggregates. Small-angle X-ray scattering (SAXS), dynamic light scattering (DLS), and static light scattering (SLS) give us important information about molecular aggregates such as their type,

K. Tsuchiya (✉)
Research Institute for Science and Technology, Tokyo University of Science, Noda, Japan
e-mail: kjtsuchi@rs.noda.tus.ac.jp

M. Abe (ed.), *Measurement Techniques and Practices of Colloid and Interface Phenomena*, https://doi.org/10.1007/978-981-13-5931-6_13

shape, and size. More reliable information about molecular aggregates can be obtained by microscopy. Molecular aggregates with sizes more than 1 μm can be directly observed with an optical microscope. However, most molecular aggregates formed by surfactants have colloid-dimensional sizes (10^{-9}–10^{-6} m); thus an electron microscope should be used to observe the structure of molecular aggregates. Since the column of a transmission electron microscope is evacuated to a very low pressure to emit an electron beam, the structure of molecular aggregates cannot be preserved in the TEM column due to the evaporation of water or oil media.

Freeze-fracture transmission electron microscopy (FF-TEM) is a useful technique for observing molecular aggregates, colloidal dispersions, and emulsions [1–7]. Rapidly frozen samples are fractured with a glass knife, and platinum and carbon are evaporated and deposited on the fractured surface to make a replica film. The replica film thus prepared is observed with a transmission electron microscope at high resolution. Because FF-TEM images depend on how molecular aggregates in the frozen sample are fractured, different images are sometimes observed even for the same molecular aggregates. FF-TEM technique can be useful to obtain the surface structure of molecular aggregates or colloidal particles (emulsions). Bilayers of liposomes (vesicles) are usually fractured at an interface between the outer monolayer (outer leaflet) and the inner monolayer (inner leaflet) in the bilayer because the interaction, van der Waals interaction, is weak. In vesicle systems, characteristic TEM images thus can be obtained depending on the conformation of hydrophobic tails of lipid molecules (or surfactants) [8–11].

13.2 What You Get

1. Observation of almost all molecular aggregates formed by amphiphilic molecules (surfactants) and colloidal particles like emulsion dispersed in either water or organic solvents (note that the image is based on the fractured surface)
2. Information about the surface structure of molecular aggregates and colloidal particles
3. Characteristic images depending on gel phase (L_β), ripple gel phase (P_β), and liquid crystal phase (L_α) for bilayers [1, 2, 8–11]

13.3 Steps and Essentials

Figure 13.1 shows a schematic illustration of the FF-TEM method [7, 11] for a typical sample, in which spherical particles are dispersed in water.

1. A small amount of the sample is rapidly frozen by plunging into liquid propane (<−170 °C). The water solvent is frozen in an amorphous state to make vitrified ice through this process.

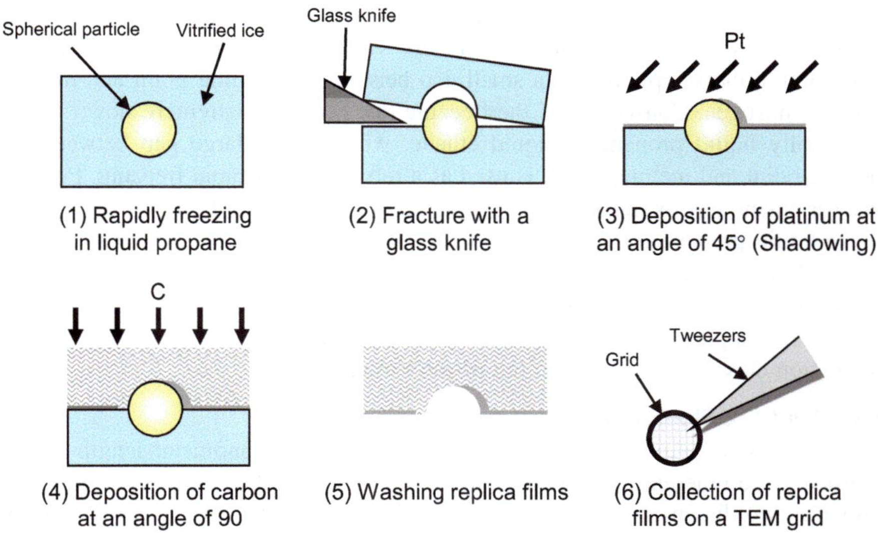

Fig. 13.1 Schematic illustration of freeze-fracture technique

2. The frozen sample is fractured with a glass knife at a low temperature (−120 to −150 °C) and under high vacuum (ca. 10^{-5} Pa).
3. Platinum (Pt) is evaporated and deposited on the fractured surface at 45° to make a thin Pt film with a thickness of ca. 3 nm. Through this process known as "shadowing," the contrast of the TEM image depending on the roughness of the fractured surface is enhanced.
4. Carbon (C) is evaporated and deposited on the fractured surface at 90° to stabilize the thin Pt film. The thickness of the C layer is typically ca. 10–20 nm.
5. The Pt/C replica film prepared with steps 3 and 4 is washed several times with solvents in which the sample can be easily soluble.
6. The replica film is placed on a TEM grid and observed with a transmission electron microscope.

Key Point (1): Rapid Freezing

The slow freezing process of water induces the nucleation of "ice crystals" and then the growth of the crystals. Ice crystals [12] must be minimized since they are artifacts and obstructive to FF-TEM observation. The formation of ice crystals stops at the recrystallization temperature (−143 °C). If the freezing speed is extremely fast, formation of ice crystals can be avoided, and amorphously frozen (vitrified) water, a structure of that cannot be observed by TEM, is formed. The rapid freezing (ideal freezing speed, >10^6 K/s) should be an important factor for FF-TEM and cryo-TEM.

How Should It Be Done?

Liquid nitrogen is for general use as a refrigerant, but it is not appropriate for the rapid freezing process since its specific heat constant is extremely low. When the sample is plunged into liquid nitrogen, the liquid nitrogen at the surface of the warmer sample is locally and immediately vaporized to make a vapor barrier surrounding the sample due to a small gap between the boiling point and melting point of nitrogen. Consequently, the freezing rate becomes extremely slow.

Usually liquid propane or liquid ethane, which has a large gap between the boiling point and melting point, is used as a refrigerant for rapid freezing. Propane gas (or ethane gas) is introduced in a bath sufficiently cooled by liquid nitrogen to make liquid propane (or liquid ethane). A small amount of the sample solution is plunged into liquid propane (or liquid ethane) by spring force for rapid freezing. An amorphously frozen region without ice crystals is limited in vicinity of the surface of the frozen sample. Then, the frozen sample should be fractured with a glass knife in the region close to the surface of the frozen sample.

Key Point (2): Shadowing

TEM can directly image the structure of the sample at nanometer length scales. However, the inherent three-dimensional (3D) structure of the sample is projected into the two-dimensional (2D) images by TEM. To solve this problem, metals (ex. platinum), which strongly scatter electrons, are evaporated and deposited onto the fractured surface at a low angle (ex. 45°). This "shadowing" process makes it possible to enhance the contrast of the roughness of the fractured surface.

Key Point (3): Washing Replica Films

The remaining of the sample on the replica films should be avoided because it causes unclear images as well as the pollution in the column of the TEM device. The sample must be completely removed out of the replica film. Usually, stirring, shaking, and ultrasonication are not used for washing the replica films because the replica films might be broken by the strong shear generated with these treatments. The replica films are kept standing in an optimized solvent until the residue of the sample is removed from the replica films. Sometimes it is difficult to select an optimized cleaning solvent and to obtain an adequate cleaning procedure especially for concentrated samples, oil-based samples, and samples containing polymers or specific surfactants (e.g., fluorocarbon-based surfactants). In most cases of dilute aqueous solutions of surfactants, acetone (or alcohol) followed by water is sufficient for washing the replica films.

13.4 Understanding Your Data

FF-TEM gives us only information about the fractured surface, not for the inherent morphology. FF-TEM images depend on how the frozen samples are fractured with a glass knife. Sometimes, it is difficult to adequately interpret the FF-TEM images. For example, FF-TEM images of oil-in-water (O/W) emulsions are similar to those of unilamellar vesicles. O/W emulsions are easily fractured at the interface between oil and water, which result in producing the hemisphere shape in the replica film. In

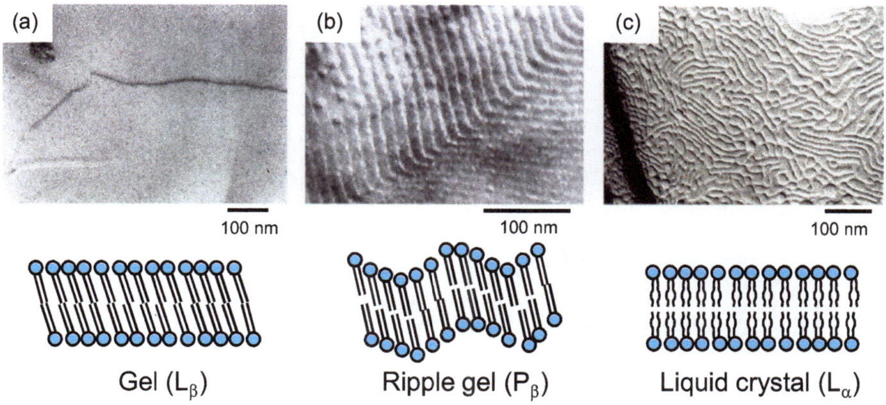

Fig. 13.2 FF-TEM micrographs of (**a**) L_β, (**b**) P_β, and (**c**) L_α phases of bilayers

the case of unilamellar vesicles, bilayer membranes are split into the outer monolayer and the inner one by the fracture process because the interaction between the outer and inner monolayers is weak. Consequently, the hemisphere shape similar to O/W emulsions is obtained in the replica films for the unilamellar vesicle. For multilamellar vesicles, information of the bilayers inside the fractured bilayer cannot be obtained.

The phase state of bilayer membranes of vesicles (liposomes) depends on temperature, and three phases, from low to high, are a gel (L_β) phase, a ripple gel (P_β) phase, and a liquid crystal (L_α) phase. Since the bilayer membrane of vesicles is fractured an interface between its outer and inner monolayers, the hydrophobic moieties are exposed on the fractured surface. The structure of the hydrophobic moieties depends on the phase state of vesicles, and characteristic images can be obtained in FF-TEM micrographs [1, 2, 8–11]. In the L_β phase, the all-*trans* conformation of the hydrocarbon chains causes a solid-like bilayer membranes of vesicles, in which amphiphilic molecules are regularly arranged. As shown in Fig. 13.2a, the "plane" structure is observed for the L_β phase. In the P_β phase, the arrangement of the amphiphilic molecules periodically changes. Then the "banded" structure is observed in the FF-TEM micrograph (Fig. 13.2b). For the L_α phase, the conformation of the hydrocarbon chain contains *trans* and *gauche*, which results in the formation of a liquid-like disordered bilayer membranes. Consequently, the "jumbled" structure is observed for the L_α phase (Fig. 13.2c).

Figure 13.3a shows an example of FF-TEM micrographs of vesicles. A larger vesicle larger than 1 μm and a smaller vesicle with the size about 200 nm are observed in this micrograph. Additionally, the shape of the fractured surfaces is the "banded" structure caused by the P_β phase. Figure 13.3b shows an example of FF-TEM of lamellar liquid crystals. In this case, the sample is fractured in the direction perpendicular to the bilayers. The width of the band-like image in this micrograph is about 100 nm and is likely to be caused by the stacked bilayers.

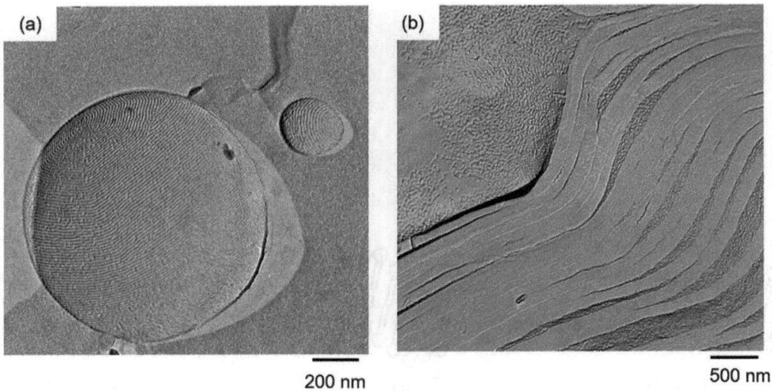

Fig. 13.3 FF-TEM micrographs of (**a**) vesicles and (**b**) lamellar liquid crystals

13.5 Useful Hints

A freeze-etching technique [13, 14] is often used together with the freeze-fracture one. These techniques are collectively called freeze-replica techniques. In the freeze-etching technique, the rapidly frozen sample is fractured with a glass knife after raising the temperature to about -100 °C and the pressure to about 10^{-4} Pa for accelerating the sublimation of the vitrified ice. This process leads to the exposure of the structure below the fractured surface. After this treatment, the replica film is formed according to the same procedure as the freeze-fracture technique. The freeze-etching technique has an advantage over the freeze-fracture technique that the height-direction contrast of the image is high.

References

1. J.A.N. Zasadzinski, S.M. Bailey, J. Electron. Microsc. Tech. **13**, 309–334 (1989)
2. W. Knoll, G. Schmidt, K. Ibel, E. Sackmann, Biochemistry **24**, 5240–5246 (1985)
3. O. Mondain-Monval, Curr. Opin. Colloid Interface Sci. **10**, 250–255 (2005)
4. Y. Shen, H. Hoffmann, J. Hao, Langmuir **25**, 10540–10547 (2009)
5. J. Hao, H. Hoffmann, K. Horbaschek, J. Phys. Chem. B **104**, 10144–10153 (2000)
6. H. Hoffmann, C. Thunig, P. Schmiedel, U. Munkert, Langmuir **10**, 3972–3981 (1994)
7. K. Tsuchiya, H. Nakanishi, H. Sakai, M. Abe, Langmuir **20**, 2117–2122 (2004)
8. B.R. Copeland, H.M. McConnel, Biochim. Biophys. Acta Biomembr. **599**, 95–109 (1980)
9. J.A.N. Zasadzinski, Biochim. Biophys. Acta Biomembr. **946**, 235–243 (1988)
10. A. Csiszar, E. Klumpp, A. Bota, K. Szegedi, Chem. Phys. Lipids **126**, 155–166 (2003)
11. K. Otake, T. Shimomura, T. Goto, T. Imura, T. Furuya, S. Yoda, Y. Takebayashi, H. Sakai, M. Abe, Langmuir **22**, 2543–2550 (2006)
12. P.M. Frederik, W.M. Busing, J. Microsc. **121**, 191–199 (1981)
13. N.M. Packter, E.R. Olukoshi, Arch. Microbiol. **164**, 420–427 (1995)
14. Y. Shang, H. Liu, Y. Hu, J.M. Prausnitz, Colloids Surf. A Physicochem. Eng. Asp. **294**, 203–211 (2007)

Chapter 14
Cryo-transmission Electron Microscopy

Koji Tsuchiya

Abstract Cryogenic transmission electron microscopy (cryo-TEM) is a valuable technique for viewing and studying the structure of various molecular aggregates such as spherical micelles, rodlike micelles, vesicles, and emulsions. In cryo-TEM, the sample under observation is usually frozen. A small amount of sample solution is placed on a grid. Excess sample is blotted away with a filter paper to form a very thin liquid film (<200 nm). This thin liquid film is rapidly plunged into a cold medium (usually liquid ethane) at temperatures lower than −170 °C. This rapid freezing induces the formation of vitrified ice, and the water molecules do not arrange into a crystalline lattice, which preserves the structure of molecular aggregates. The freezing rate is one of the most sensitive parameters for determining the quality of the structure preservation. The frozen sample is transferred to a cryospecimen holder using a cryo-transfer system under a liquid nitrogen atmosphere and is observed directly by TEM while the sample is kept in the frozen state.

Keywords Cryo-TEM · Direct imaging · Rapid freezing · Vitrified ice

14.1 Introduction

Transmission electron microscopy (TEM) is one of the most useful tools since it gives us information about the morphology of various self-assemblies formed by surfactants. As shown in Chap. 13 (FF-TEM), because liquid samples cannot be directly loaded into the column of a TEM device under vacuum, adequate pretreatment is required. Since surfactant molecules mainly consist of light elements such as carbon (C), hydrogen (H), oxygen (O), and nitrogen (N), electrons cannot be sufficiently scattered, which results in the low contrast of TEM images. Negative staining [1–3] is an established and early method to improve this problem. Some heavy metal salts such as uranyl acetate and phosphotungstic acid are chosen as the

K. Tsuchiya (✉)
Research Institute for Science and Technology, Tokyo University of Science, Noda, Japan
e-mail: kjtsuchi@rs.noda.tus.ac.jp

M. Abe (ed.), *Measurement Techniques and Practices of Colloid and Interface Phenomena*, https://doi.org/10.1007/978-981-13-5931-6_14

stain because they can scatter electrons strongly. A small amount of the sample is deposited on a TEM grid, and the sample is covered with a small amount of the stain. Excess sample and stain are blotted away with a filter paper. After drying, the grid obtained is observed by TEM [1–3]. Since the electron beam passes through the sample but is highly scattered at the stain surrounding the sample, a "negative" image with high contrast is obtained. However, the drying and staining processes cause a shape change and collapse of the sample. Consequently, the negative staining image might be totally different from the original structure of the sample. In the case of FF-TEM described in Chap. 13, the drying and staining processes are not required. However, it is difficult for us to understand the overview image of the structure because the FF-TEM image gives us an only information about the fractured surface of the sample. How should we observe molecular aggregates such as micelles and vesicles without changing the original structure as much as possible? Cryo-transmission electron microscopy (cryo-TEM) is the most useful and prominent technique to obtain the direct image of the structure of molecular aggregates [4, 5]. For cryo-TEM, molecular aggregates formed in the sample are embedded in vitrified ice formed by rapid freezing and are observed directly by TEM while the sample is kept in the frozen state.

14.2 What You Get

Direct observation of the structure of molecular aggregates at high resolution without staining and drying processes.

14.3 Steps and Essentials

Figure 14.1 shows a schematic illustration of a cryo-TEM method [6, 7].

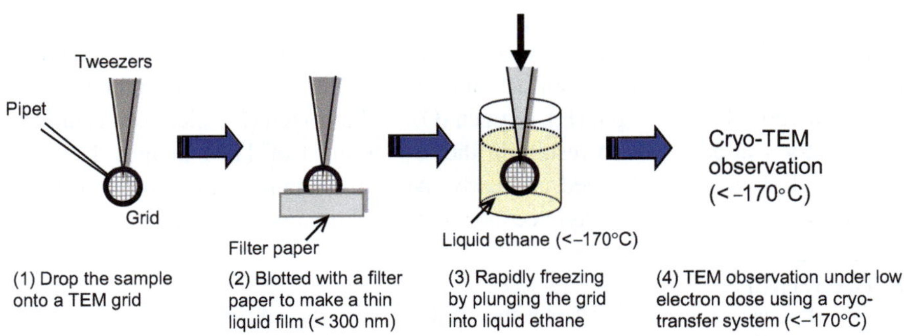

Fig. 14.1 Schematic illustration of the procedure for cryo-TEM

1. A TEM copper grid covered by a holey carbon film is changed into hydrophilicity by a glow discharge treatment, which allows aqueous samples to spread easily on its surface. A small amount (3–5 μL) of sample solution is placed on the grid, which is held by a pair of self-locking tweezers mounted on a spring-loaded shaft with a cryo-preparation system.
2. The sample drop is blotted with a filter paper to form a thin liquid film (<300 nm) on the grid.
3. Immediately after making the thin liquid film, the grid is plunged into liquid ethane (<−170 °C) sufficiently cooled by liquid nitrogen.
4. The grid is transferred onto the tip of a cryospecimen holder using a cryo-transfer system under a liquid nitrogen atmosphere. The rapidly frozen sample is kept at temperatures lower than −170 °C and imaged in a transmission electron microscope under low electron dose.

As shown above, complicated multiprocesses are performed for cryo-TEM observation. There are many points to which attention should be paid. In particular, the following points are important to obtain a good image of cryo-TEM.

Key Point (1): Freezing Speed

When the freezing rate of water is slow, ice crystals, which should be artifacts, are formed [8]. Thus, the rapid freezing is an important point for cryo-TEM observation (see Chap. 13 for details). In general, liquid propane and liquid ethane are used as a refrigerant for rapid freezing. In the major cases of cryo-TEM, liquid ethane is used because liquid ethane is sublimed and removed from the frozen sample more easily than liquid propane. After making a thin liquid film of the sample, it is plunged into liquid ethane using a spring for rapid freezing.

Key Point (2): Thickness of the Vitrified Sample

In transmission electron microscopy, the electrons go through the specimen to obtain an image. Thus, the thickness of the sample must be decreased. For cryo-TEM, excess sample is blotted with a filter paper to make a thin liquid film, which is then vitrified by plunging into liquid ethane. If the vitrified sample is too thick, the following problems occur [1]. Because the electrons cannot go through the sample, cryo-TEM images cannot be obtained [2]. The freezing rate decreases toward the inside of the sample. Then, ice crystals are easily formed in the thick sample [3]. The thick vitrified sample is seriously damaged by the electron beams.

On the other hand, when the sample is too thin, other following problems occur [1]. The vitrified sample film is collapsed by the electron beams because the strength of the sample is too weak [2]. Large molecular aggregates cannot be embedded in the vitrified ice. Thus, the optimization of the thickness of the vitrified sample should be the most important point to obtain a cryo-TEM image.

The blotting speed of the sample with a filter paper varies depending on various physical properties of the sample (e.g., viscosity). In addition, we cannot perform this blotting process while measuring the thickness of the liquid film. Thus, it is sometimes necessary to repeat trial and error in order to obtain a thin film having an appropriate thickness.

Key Point (3): Electron Beam Damage

Organic compounds such as surfactants, polymers, and biological materials are electron beam-sensitive materials compared to metals and solid particles. These materials are seriously damaged by electrons, resulting in a morphological change due to heating and electrostatic charging. Thus, for cryo-TEM, the sample has to be observed in the shortest time possible, while the electron beam dose is extremely low. In fact, cryo-TEM images are just slightly observed on the fluorescent screen of the TEM device. Usually, the sample is imaged with a high-resolution CCD camera for cryo-TEM. The cryo-TEM observation should be performed using the low-dose condition mode installed in a TEM device. A smaller spot size of the electron beam is better as the electrons are not irradiated to the virgin region. Cryo-TEM images have low scattering contrast because the vitrified sample mainly consisting of low elements is observed without staining. The most common approach to generate contrast in cryo-TEM is to record images with a slight deviation from just-focus to under-focus, that is, the phase-contrast (defocus) imaging.

Cryo-TEM is a useful technique for directly imaging of molecular aggregates formed in water. However, applicable samples are limited. The following samples are unsuitable for cryo-TEM:

1. Oil-continuous system (e.g., reversed micelles, reversed vesicles, and water-in-oil (W/O) emulsions)

Frozen oils are more seriously damaged by electrons than vitrified water. There are some reports about oil-continuous systems [9, 10], but it is basically difficult or sometimes impossible to apply cryo-TEM to oil-continuous systems (especially to alkane-based systems).

2. Samples containing large molecular aggregates

For cryo-TEM, a sample thinned to a thickness that allows electron beam penetration is rapidly frozen, and molecular aggregates embedded in the amorphous ice are observed. Thus, it is not suitable for observation of molecular aggregates larger than the thickness of the thin film. For example, when the sample is observed at an accelerating voltage of 100 kV, the thickness of the sample that electron beam can go through is about 200–300 nm. Larger molecular aggregates might not be retained in the thin film or might be deformed by stress caused by the thinning process. Even if molecular assemblies larger than 300 nm are observed, their shapes and sizes might be different from the original ones; thus, they cannot be evaluated correctly.

3. Concentrated samples

Concentrated samples are unsuitable for cryo-TEM because molecular aggregates appear to overlap in the image. Additionally, the electron beam damage is serious as compared to the dilute system.

4. High viscous samples

Samples with high viscosity are possible, but it is difficult to observe by cryo-TEM. It is often impossible to adequately remove the excessive amount of the sample with a filter paper to make a thin film.

As for concentrated samples [3] and high viscous samples [4], if the association form does not change by dilution, it is better to observe for diluted samples.

14.4 Understanding Your Data

Cryo-TEM images should be taken in the area without ice crystals, which are artifacts formed in the slow-freezing region. In the cryo-TEM image of Fig. 14.2a, spherical particles (arrows in the figure) having high contrast are observed in addition to unilamellar vesicles (ring-shaped image). These spherical particles are due to ice crystals. It is difficult to distinguish small oil droplets and polymer globules from ice crystals in cryo-TEM images.

The white spotlike patterns of Fig. 14.2b are due to holes caused by electron beam (electron beam damage). Since the holes caused by electron beam damage expand depending on the irradiation time of electron beam, they can be identified easily. In such a case, it is necessary to observe by further reducing the irradiation amount of the electron beam or to observe the sample in a short time.

Figure 14.3a and b show typical cryo-TEM images of wormlike micelles [11, 12] and vesicles [6 7, 13], respectively. In Fig. 14.3a, wormlike micelles with diameters of 4–5 nm and lengths in micrometer range are entangled with each other to form three-dimensional networks. In Fig. 14.3b, unilamellar vesicles and some linear structures (arrows in the figure) are observed. The linear images are not due to rodlike micelles. If they are rodlike micelles, the contrast is much smaller, and the diameter of the line should be smaller (compare them with the image of wormlike

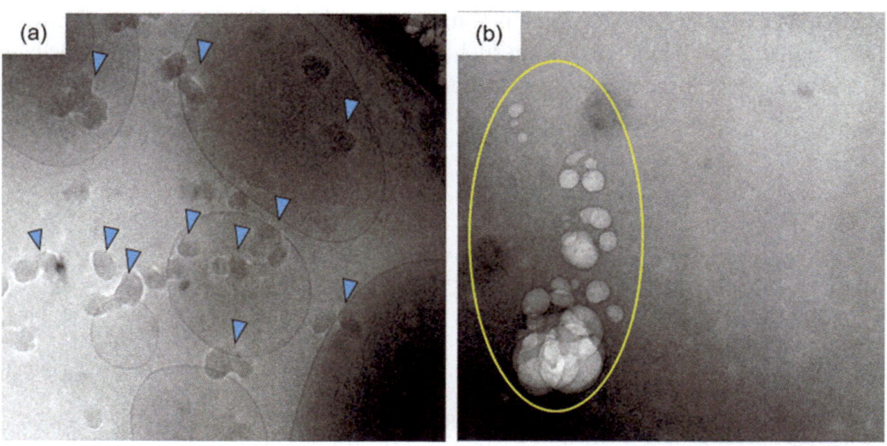

Fig. 14.2 Cryo-TEM micrographs (**a**) including ice crystals and (**b**) damaged by electron beam

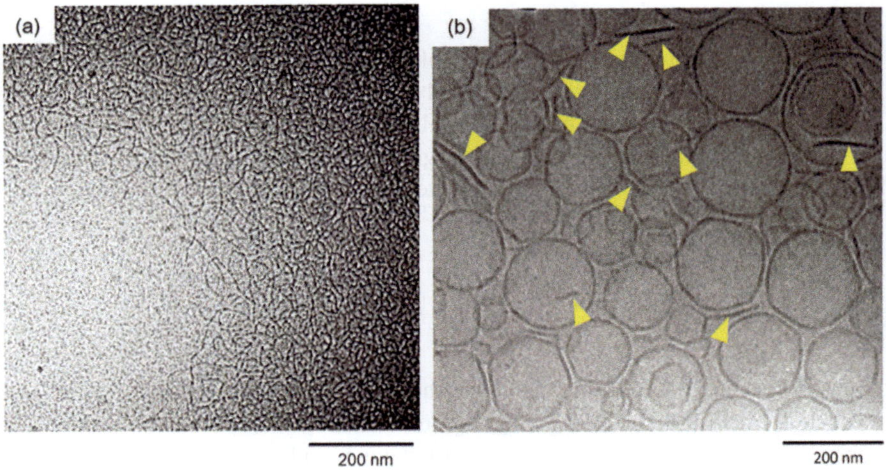

Fig. 14.3 Cryo-TEM micrographs of (**a**) wormlike micelles and (**b**) vesicles

micelles in Fig. 14.3a). These linear shapes are likely to be from an image observing a lamellar disc dispersed in water from the side. When observing the lamellar disc from the top, there is hardly any contrast, but when observing from the side, a clear linear contrast is obtained. Linear images due to such lamellar discs are often observed in vesicle dispersion systems.

14.5 Useful Hints

Since cryo-TEM is not subjected to dyeing treatment and the electrons are directly irradiated to the vitrified sample, the contrast of the image is very low and electron beam damage is intense. It is often difficult to obtain desired cryo-TEM images. In such a case, cryo-negative staining is sometimes performed instead of cryo-TEM. In this method, a small amount of stain such as phosphotungstic acid or uranium acetate is added to the sample to improve the contrast of cryo-TEM images.

References

1. M. Maskos, J.R. Harris, Macromol. Rapid Commun. **22**, 271–273 (2001)
2. M.F. Ottaviani, P. Matteini, M. Brustolon, N.J. Turro, S. Jockusch, D.A. Tomalia, J. Phys. Chem. B **102**, 6029–6039 (1998)
3. T. Imura, Y. Tsukui, T. Taira, K. Aburai, K. Sakai, H. Sakai, M. Abe, D. Kitamoto, Langmuir **30**, 4752–4759 (2014)
4. Y. Talmon, Surfactant Sci. Ser. **83**, 147–178 (1999)
5. V. Alfredsson, Curr. Opin. Colloid Interface Sci. **10**, 269–273 (2005)

6. H. Sakai, T. Saitoh, T. Misono, K. Tsuchiya, K. Sakai, M. Abe, J. Oleo Sci. **60**, 563–567 (2011)
7. K. Tsuchiya, J. Ishikake, T.S. Kim, T. Ohkubo, H. Sakai, M. Abe, J. Colloid Interface Sci. **312**, 139–145 (2007)
8. P.M. Frederik, W.M. Busing, J. Microsc. **121**, 191–199 (1981)
9. D. Danino, R. Gupta, J. Satyavolu, Y. Talmon, J. Colloid Interface Sci. **249**, 180–186 (2002)
10. L. Wolf, H. Hoffmann, Y. Talmon, T. Teshigawara, K. Watanabe, Soft Matter **6**, 5367–5374 (2010)
11. Y.I. Gonzalez, E.W. Kaler, Curr. Opin. Colloid Interface Sci. **10**, 256–260 (2005)
12. A. Bernheim-Groswasser, R. Zana, Y. Talmon, J. Phys. Chem. B **104**, 4005–4009 (2000)
13. K.L. Herrington, E.W. Kaler, D.D. Miller, J.A. Zasadzinski, S. Chiruvolu, J. Phys. Chem. **97**, 13792–13802 (1993)

Chapter 15
Zeta (ζ) Potential for Micelle and Microemulsion

Masahiko Abe

Abstract Measurement of zeta potential is not applicable to systems with entangled particles, non-charged particles, and concentrated solutions where the interfacial electrical double layers are depressed. However, dispersibility of charged colloidal particles in aqueous solution can be evaluated by ζ potential. Therefore, this chapter deals with the basics of electrical interface phenomenon and its relationship between surface electric potential, zeta potential, and surface charge density.

Keywords ζ potential · Charged fine particles · Electrophoretic light scattering · DLVO theory

15.1 Introduction

Measurement of ζ potential is commonly applied to evaluate the dispersibility of charged solid or liquid fine particles in aqueous electrolyte solutions. As the obtained ζ potential itself would not represent the effect of particle size, unit surface charge should be calculated for the evaluation of the object particle. Unit surface charge can be determined from the ζ potential divided by its surface area.

15.2 What Can Be Known from ζ Potential Measurement

Dispersibility of charged colloidal particles in aqueous solutions can be evaluated by the ζ potential. This method is not applicable to systems with entangled particles, non-charged particles, and concentrated solutions where the interfacial electrical double layers are depressed.

M. Abe (✉)
Research Institute for Science and Technology, Tokyo University of Science, Noda, Japan
e-mail: abemasa@rs.noda.tus.ac.jp

M. Abe (ed.), *Measurement Techniques and Practices of Colloid and Interface Phenomena*, https://doi.org/10.1007/978-981-13-5931-6_15

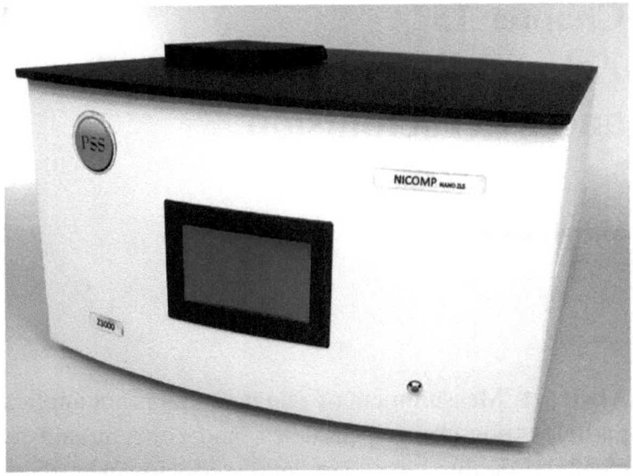

Fig. 15.1 Zeta potential apparatus Nicomp Nano 3000 ZLS (Photo credit Nippon Entegris K.K)

15.3 Measurement of Zeta (ζ) Potential

Many commercial instruments are available for the measurement of zeta (ζ) potential, and the instruction manual should be referred to for actual measurement. Here, we explain theoretical background and measurements following Nicomp Nano 3000 ZLS (PSS Japan) which we use in our laboratory (Fig. 15.1). The characteristic features of Nicomp Nano 3000 ZLS are its capability for low electric field, reduction against Joule heat effect caused by random movement, short time measurement, high reproducibility and reliability, applicability for nonaqueous solution, and compensational function for the charge polarization indispensable to measure highly salted samples. Figure 15.2 shows the basic principle of the Nicomp system, and Fig. 15.3 shows Nicomp's dip cell.

The procedure for measurement is as follows, and data will be shown automatically.

1. Place the sample into the cell shown in Fig.15.3 (left).
2. Set the electrode as shown in Fig.15.3 (right).
3. Select PALS mode shown on the PC monitor.
4. Set the electric field level (normally 5 v/cm or less, c.a. 2 v/cm).
5. Input the viscosity of the solvent (0.8922 for water under 25 °C) and the reflective index of the solvent (1.33287 for water under 25 °C).

15.4 Important Point for the Measurement

The basic principle of this instrument for ζ potential measurement is electrophoretic light scattering (ELS) with two analytical modes, i.e. frequency analysis and phase analysis. There are limitations in frequency analysis for particles with slow mobility because this method is based on the phase shift of frequency spectrum between the

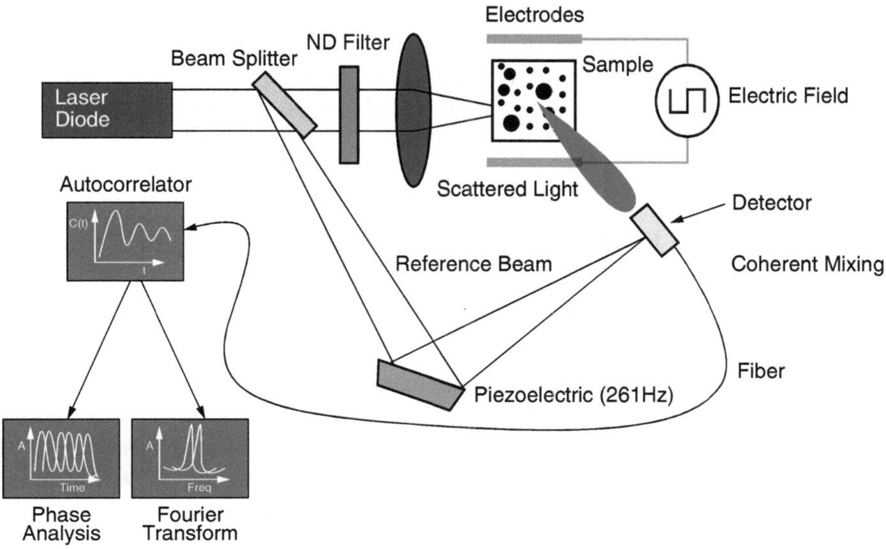

Fig. 15.2 Measurement principle of Nicomp zeta potential analyser

Fig. 15.3 Electrode cell

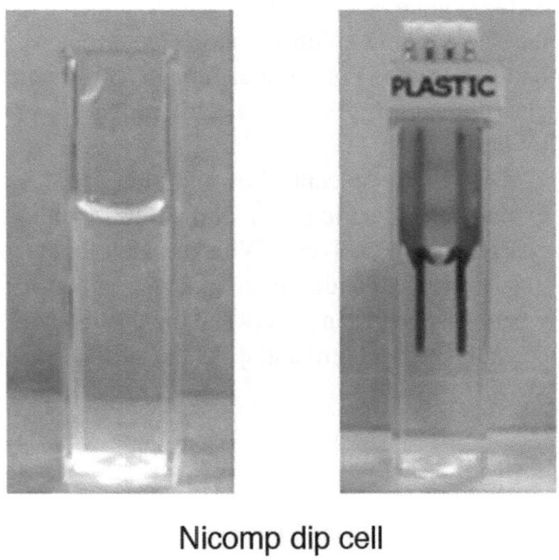

particles with and without applied electric field. For example, lower electric field level is preferable to suppress Joule heat and resulted convection heat, but the phase shift will be suppressed, and the S/N ratio lowers and will lose the accuracy of measurement. As a result, it is difficult to measure middle to highly salted samples or organic solvents with a dielectric constant much lower than water, because the particle mobility under electrophoresis is significantly small in both cases.

The value of phase analysis should be noted for these cases, as this method is based on the phase shift from the Doppler effect of the positional change of the particles under electric field application by electrophoresis. This method is insensible to Brownian movement and enables high precision measurement.

15.5 Basics of Electrical Interface Phenomenon [1]

15.5.1 Interfacial Electrical Double Layers

Due to the active Brownian movement in colloidal solutions, coagulation and sedimentation of particles are barely observed under long-term observation despite the frequent collision of particles. Generally, colloidal particles tend to be charged by many causes, such as absorption of ions and electric dissociations. Under different dielectric constants (Dc) between particles and dispersant, larger Dc charges positive, and smaller Dc charges negative. As the total system should remain neutral as a whole, electric charges on particle surfaces have uneven distribution as designated to electrical double layers (EDL). The electric potential difference caused by EDL is called double layer potential (DLP), which exists both on the particle surface and in the surrounding solution, and is determined by the concentration of ions mobile between them. For dissociation of carboxylic groups on the surface, H^+ is the determinant ion. As such, electric properties of solid/liquid interfaces are controlled by DLP.

This EDL concept shown in Fig. 15.4 was proposed by Helmholtz, which explains the electric potential distribution between a flat solid surface with positive charge and the solvent. With the Helmholtz model, EDL can be considered as a planar parallel condenser so that charge density (σ) can be expressed by Eq. (15.1), where ζ is the distance between the positive and negative charges, ε_0 is the dielectric constant for vacuum, and φ_0 is the surface electric potential.

$$\sigma = \frac{\varepsilon_r \varepsilon_0 \phi_0}{\zeta} \tag{15.1}$$

In actuality, electric potential linearly decreases from the solid surface to the bulk solution to become zero, and ions in the solution are attracted to the opposite-charged solid surface while tending to dislocate due to thermal motion at the same time. As a result, the distribution of ions in the EDL is diffusive, determined by the balance between electrostatic attraction and thermal motion. Gouy and Chapman independently proposed diffuse EDL models as shown in Fig. 15.5 with its potential charge profile. Ions with charges opposite to the solid surface distribute widely in the solution, and the average distance from the surface ($1/\kappa$) can be determined by Eq. (15.2) where F is Faraday constant, ε is dielectric constant of the solution, R is Gas constant, T is absolute temperature, and J is ionic strength.

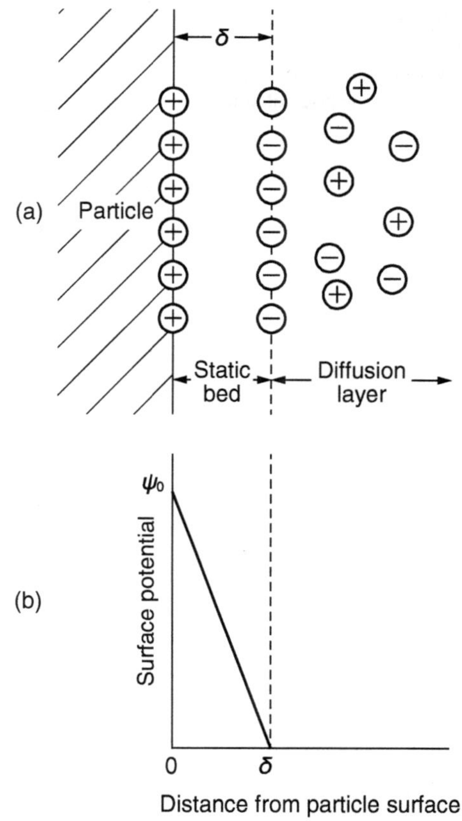

Fig. 15.4 Relationship between electrical double layer of Helmholtz model and change of electric potential

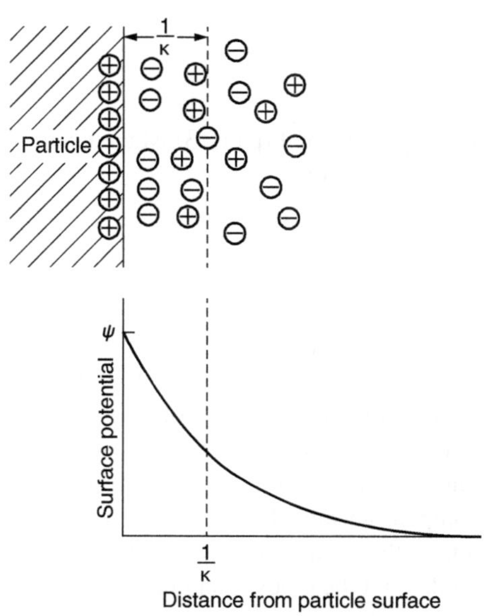

Fig. 15.5 Relationship between electrical double layer of Gouy-Chapman model and change of electric potential

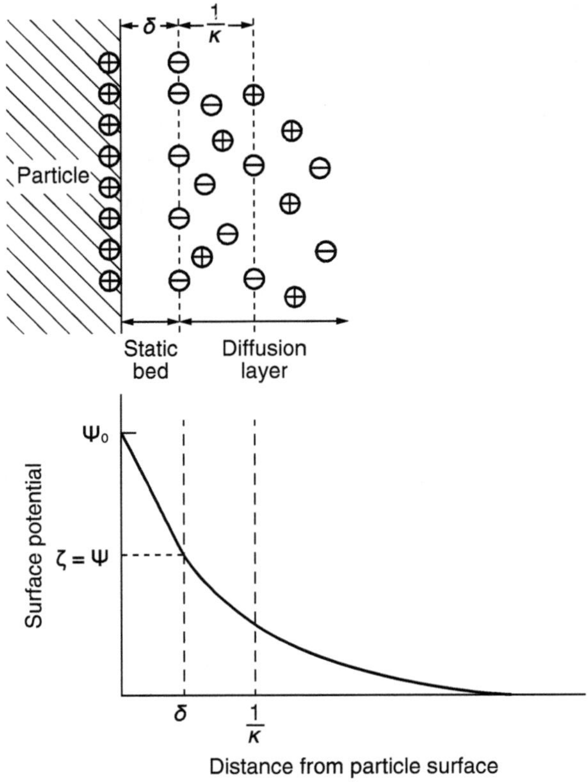

Fig. 15.6 Relationship between electrical double layer of Stern model and change of electric potential

$$\kappa = \frac{\left(8\pi F^2\right)^{\frac{1}{2}}}{(1000\varepsilon RT)^{\frac{1}{2}}(J)^{\frac{1}{2}}} \tag{15.2}$$

J is expressed by Eq. (15.3) where *Ci* and *Zi* correspond to the concentration and ion valence of ion *i*.

$$J = \frac{\sum C_i Z_i^2}{2} \tag{15.3}$$

Increase in the ionic strength by addition of electrolytes in the colloidal solution reduces the thickness ($1/\kappa$) of EDL from Eqs. (15.2 and 15.3). These equations indicate that the thickness of EDL becomes all the same regardless of the species of ions if the valence of all the ions are equal. In actuality, there are differences in EDL thickness depending on the type of ions. Stern proposed another EDL model to compensate such discrepancies as shown in Fig. 15.6. This model has two layers, an internally fixed Stern layer and a diffuse layer, as a combination of the Helmholtz model and the Gouy-Chapman model. Part of the ions adhere to the Stern layer, and the rest of the ions disperse in the diffuse layer to give a linear potential decrease in

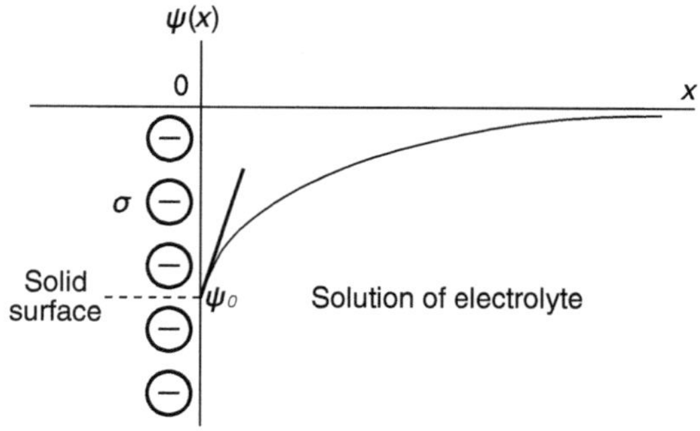

Fig. 15.7 Relationship between potential of solid surface and electrical potential distribution

the Stern layer and then exponentially decay in the diffuse layer towards the bulk solution as shown in Fig. 15.6. ζ potential is the electric potential at the plane in between the Stern layer and the diffuse layer called the slipping plane which can be experimentally measured. When the thickness of EDL is negligibly small, it can be equal to the surface potential Ψ. ζ potential correlates to the surface charge density σ and the thickness of EDL ($1/\kappa$) as shown in Eq. (15.4), and ζ increases with the increase of surface charge density or decrease of EDL thickness. ζ potential gives a good indication of the dispersion stability, though DLVO theory is often used for strict discussions.

$$\zeta = \frac{\sigma}{\varepsilon_r \varepsilon_0 \kappa} \tag{15.4}$$

15.5.2 Relationship Between Surface Electric Potential, Z(ζ) Potential, and Surface Charge Density

Figure 15.7 shows electric profiles of charged solid surfaces in electrolyte solutions, where x expresses the distance in the solution from the solid surface and Ψ(x) shows the electric potential of which bulk potential is set to 0 ($\Psi(\infty) = 0$). The solid surface charge is regarded as a constant.

There are two potentials at x = 0. One is $\frac{d\psi}{dx}\big|_{x=+0}$, shown as a tangent thick line in Fig. 15.7, which is the right-side limit value corresponding to the approach from the bulk side which is positive. The other is $\frac{d\psi}{dx}\big|_{x=-0}$ in which the potential value is 0. We take the right-side limit value here for further discussion. Ψ_0 (v) expresses the solid surface potential, and σ (C/m^2) is the surface electric density where the electrolyte is a symmetric type with valence value ν.

$$\left.\frac{d\psi}{dx}\right|_{x=+0} = -\frac{\sigma}{\varepsilon_r \varepsilon_0} \quad (15.5)$$

$$\frac{d^2\psi}{dx^2} = -\frac{\text{ven}}{\varepsilon_r \varepsilon_0}\left[e^{-}\frac{\iota e\psi(x)}{kT} - e\frac{\text{v}e\psi(x)}{kT}\right] \quad (15.6)$$

$$\sigma = \sqrt{8CN_A\varepsilon_r\varepsilon_0 kTx10^3}\sinh\left(\frac{\nu e\psi_0}{2kT}\right) \quad (15.7)$$

where C is concentration (mol/l), N_A is Avogadro's number, *e* is elementary charge (e = 1.602 × 10^{-19} C), $\boldsymbol{\varepsilon_r}$ is dielectric constant for water (78.5 at 25 °C), $\boldsymbol{\varepsilon_0}$ is dielectric constant for vacuum (78.854 × 10^{-12}F/m), n is concentration of electrolyte (l/m^3), ***T*** is absolute temperature (K), and ***k*** is Boltzmann constant (1.38 × 10^{-23} J/K).

When valence value of the electrolyte is 1 ($\nu = 1$) and surface potential Ψ_0 (v) is expressed by zeta potential (ζ_0), Eq. (15.7) can be applied to calculate the surface electric potential σ (C/m^2) of the dispersant such as micelles in colloidal solutions by measuring (ζ) vs the concentration of the surfactant.

15.6 Calculation of Surface Electric Potential from the Measured Zeta (ζ) Potential

The zeta (ζ) potential and the particle size of microemulsions consisting of sodium dodecyl-sulphate in 0.086 M NaCl/n-octane/n-hexane are measured at 30 °C. The surface electric potential σ (C/m^2) is calculated by Eq. (15.7) with a valence value 1 as shown in Table 15.1, and both the ζ potential and the particle size are measured.

Table 15.1 Particle sizes and zeta potentials of microemulsion

Concentration of electrolytes (mol/l)	Particle size (diameter) (nm)	Zeta potential (mV)	Surface charge density (C/m^2)
0.08	632	−46.7	①
0.086	34.4	−38.5	②
0.086	48.4	−39.7	③
0.086	55.3	−32.1	④
0.086	61.7	−31.0	⑤

15.7 One-Point Merit for the Determination of Surface Electric Potential

The surface electric potential of particles at a specific concentration and temperature can be determined by measuring the fluid dynamic diameter of particles with dynamic light scattering, which is converted to the surface area. The surface electric potential can be obtained from the surface area by dividing the zeta (ζ) potential as shown in Eq. (15.8), where ζ is zeta potential of the object particle and $\boldsymbol{r}$ is its radius.

$$\zeta/4\pi r^2 \tag{15.8}$$

Reference

1. H. Ohshima, *Theoretical of Colloid and Interfacial Electric Phenomena*, vol 12, 1st edn. (Academic Press, San Diego, 2006)

Chapter 16
Electron Microscopy Observation of Solid Particles

Kanjiro Torigoe

Abstract Electron microscopy is an indispensable tool to observe and analyze physical characteristics of small samples being invisible under visible light. In this chapter, first the reason why the electron microscopy is utilized is briefly explained. Subsequently basic structures of scanning electron microscopy (SEM) and transmission electron microscopy (TEM) are presented. Finally, answers for practical questions, e.g., how to prepare the samples and what is the appropriate image, will be given.

Keywords SEM · TEM · Bright-field image · Dark-field image · Electron diffraction

16.1 Introduction

What does it mean that an object is visible? For example, supposing two small objects are placed at a certain distance apart. You can recognize the two when the distance is sufficiently large. But if this distance is too short, they might be recognized as a single object, and then we cannot say that the objects were properly observed. Here "short" and "large" are compared with the resolution (in precise, the point resolution) of your apparatus. The resolution of optical lenses is defined as follows:

$$\delta = \frac{0.61\lambda}{\mathrm{NA}} \tag{16.1}$$

where λ is the wavelength of light source and NA refers to the number of aperture of the optical lens, normally NA < 1. From Eq. (16.1), one finds that for a conventional optical microscope using visible light (λ = 380–700 nm) and NA = 0.8, the

K. Torigoe (✉)
Department of Pure and Applied Chemistry, Tokyo University of Science, Noda, Japan
e-mail: ktorigoe@rs.noda.tus.ac.jp

M. Abe (ed.), *Measurement Techniques and Practices of Colloid and Interface Phenomena*, https://doi.org/10.1007/978-981-13-5931-6_16

resolution is 290 nm, below which one cannot see. To observe an object with a smaller size, we use microscopes bearing a light source with smaller wavelengths (i.e., electron microscope) or other microscopes utilizing tunneling current (scanning tunneling microscope, STM) or force (atomic force microscope, AFM) acting between the probe and the specimen.

Needless to say, the electron microscope is an equipment to observe objects of very small size (typically less than 1 μm) using an accelerated electron beam as light source. Since the accelerated electrons have a wavelength much shorter than that of visible light, the electron microscope allows us to observe very small objects that cannot be observed with optical microscope. The wavelength of accelerated electrons is given by the following equation:

$$\lambda = \frac{h}{\sqrt{2m_0\mathrm{eE}}} \quad \text{(without correction by relativistic effect)} \tag{16.2}$$

or

$$\lambda = \frac{h}{\sqrt{2m_0\mathrm{eE}\left(1+\frac{\mathrm{eE}}{2m_0c^2}\right)}} \quad \text{(corrected with relativistic effect)} \tag{16.3}$$

where h is Planck constant (6.626×10^{-34} J s), m_0 denotes static mass of electron (9.109×10^{-31} kg), e is the electronic charge (1.602×10^{-19} C), E is the accelerated voltage, and c is the light velocity (2.998×10^{8} m). Correction due to the relativistic effect is necessary when electrons are accelerated at greater than 100 kV, where the speed of electrons approaches the light velocity and the increase in the mass becomes not negligible.

The electron microscopes are classified into two categories: scanning electron microscope (SEM) and transmission electron microscope (TEM). The typical values of accelerated voltage are 10 kV and 200 kV for SEM and TEM, respectively. When these values are applied to Eqs. (16.2) and (16.3), respectively, the wavelengths of electrons are calculated to be 12.3 pm and 2.5 pm (1 pm = 10^{-12} m). Supposing that the atomic diameter is ca. 0.2 nm (200 pm), one finds that individual atoms can be observed with a TEM at 200 kV of accelerated voltage, if electromagnetic lenses have ideal properties.

16.2 What You Get

16.2.1 Scanning Electron Microscopy (SEM)

Scanning electron microscopy (SEM) provides information on surface morphology for a micrometer or sub-micrometer region of the samples. The model structure of typical SEM is illustrated in Fig. 16.1.

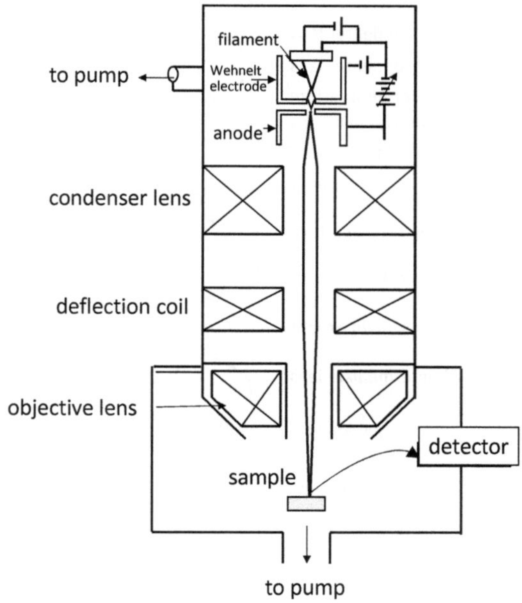

Fig. 16.1 Model structure of SEM

In SEM, accelerated electrons at a comparatively low voltage (< 10 kV) are irradiated to the sample by scanning the position two-dimensionally. Then both electrons emitted from the interior of the sample (secondary electrons) and those undergone Compton scattering by the electrons within the sample (recoil electrons) contribute to make an image. Here, the secondary electrons have energy of several tens of eV, irrespective of the energy of incident electrons, and generation depth is ca. 10 nm. The emission efficiency is dependent on the tilt angle of the sample; greater emission efficiency can be obtained at larger tilt angle, providing a high-contrast image. On the other hand, the energy and the generation depth of recoil electrons are dependent on the energy of incident electrons. An image obtained at higher acceleration energy contains information on the sample composition at a greater depth.

Generally, the sample for SEM is prepared as follows. First, a conductive tape (i.e., carbon tape) is adhered to the top surface of specimen holder, and the sample is deposited on it. Then the sample holder is placed in a vapor deposition apparatus, and a heavy metal (Pt, Pd, Au) is vapor-deposited prior to the observation. The heavy metal contributes to increase emission of recoil electrons to improve the image contrast. However, the vapor deposition cannot be conducted when you do elemental analysis by energy-dispersive X-ray (EDX) associated with your SEM.

16.2.2 Transmission Electron Microscopy (TEM)

In TEM, accelerated electrons at a high voltage (100–300 kV) were irradiated to the sample. The electrons transmitted from the sample and a part of scattered electrons

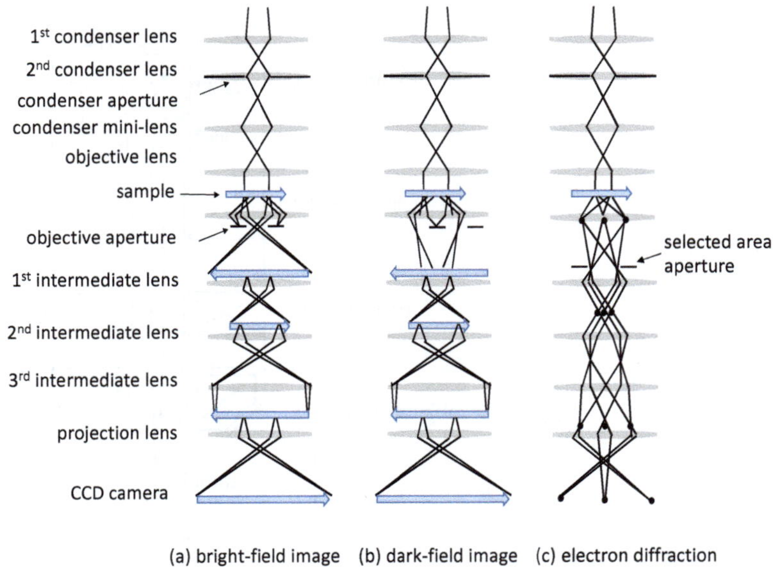

Fig. 16.2 Light itineraries for different modes of TEM

are used for making image. The TEM comprises three systems: (1) an emission system to draw electrons from a light source (typically W or LaB_6) and accelerate them, (2) an imaging system to enlarge the sample image and attach a high contrast to it, and (3) a camera system to record images. Both the emission and imaging systems include several electromagnetic lenses. Magnification of the image is determined by varying electric current of the lenses in the imaging systems.

The TEM has three different observation modes, as explained below (Fig. 16.2):

1. Bright-Field Image

The bright-field image is the most general observation mode in TEM. Here we can obtain information on size and shape of the sample (typically <100 nm in size). When accelerated electrons are irradiated to the sample, some electrons are scattered, and the others are transmitted. The fraction of scattered electrons is proportional to the atomic number Z of the sample. The scattered electrons with large scattering angles are blocked by the objective aperture, while those with small scattering angles and transmitted electrons contribute to forming the image (scattering contrast). On the other hand, unlike SEM, little information is available on the surface morphology, yet some weak contrast appears due to interference between the transmitted and small-angle scattered electrons (diffraction contrast or phase contrast).

Note that different images are recorded depending on the focused position. When the electron beam is focused above the screen of CCD camera, a high-contrast image with white fringes is obtained along the object (under-focus image, Fig. 16.3a). Conversely, when the electron beam is focused below the CCD, an image with black fringes appears (over-focus image, Fig. 16.3c). When the electron beam is focused at an appropriate position, the fringe is absent, but here the image contrast becomes

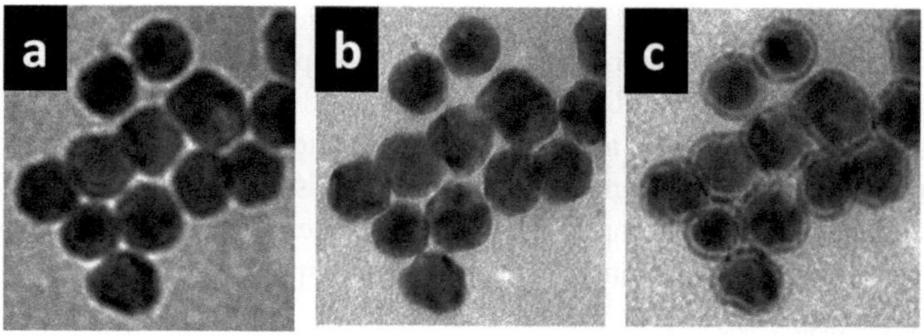

Fig. 16.3 Variation of TEM images by focus position of the electron beam. (**a**) Under-focus image, (**b**) appropriate-focus image, (**c**) over-focus image

minimum (appropriate-focus image, Fig. 16.3b), so it might be difficult to image light element samples or very small particles at the appropriate focus. In such a case, it is better to record the image at a slightly under-focus position.

The way to adjust focus position might be slightly different for different electron microscopes. However in general, after your sample is introduced in the TEM, first you adjust coarsely the height of your sample by twisting a screw below the sample chamber, and then you adjust the focus finely by using a focus adjust and a wobbler. The focal point is appropriate when the shake of image by applying wobbler becomes minimum.

For TEM observation, the sample must be thinner than ca. 100 nm to transmit electrons; therefore pre-treatment might be required depending on the sample. For example, for bulk metal sample, chemical etching and electric etching are applied for thinning edge of the sample. For powder samples, first it is dispersed in a proper solvent, under sonication if necessary, and a portion of the dispersion is collected with a Pasteur pipette or a micropipette.

Then it is mounted on a TEM grid. The TEM grid has circular form with 3 mm diameter, mostly made of copper and covered with polymer membrane (nitrocellulose (collodion), formvar) and reinforced with vapor-deposited carbon. Subsequently the sample grids are well dried in a desiccator prior to the observation. Collodion-attached carbon grids are commercially available.

When the sample comprises exclusively light elements, it might happen that the image contrast is so low that we cannot judge whether the focus is appropriate or not. In such a case, negative staining would be useful for the observation. For negative staining, samples are immersed in a dilute (1–2 wt%) staining solution dissolving heavy metal ions, i.e., phosphotungstic acid, osmium tetroxide, and uranyl acetate. Some of these molecules or ions adsorb on the sample surface, and others remain in the bulk solution. Consequently, all phases except for the inside of the sample are stained by heavy metal atoms and thus look dark. Other techniques, involving freeze fracture and cryogenic TEM, are explained in later chapters.

2. Dark-Field Image

The dark-field image is derived from scattered electrons by shielding transmitted electrons with the objective aperture. This method is useful for discerning different domains composed of different elements and/or different crystal facets. This method is accompanied with selected area electron diffraction (SAED). For that, first you observe a bright-field image of the sample at the appropriate focus, and then you change the mode to electron diffraction. Subsequently you hide all diffraction spots on the fluorescent screen with a selected area diffraction aperture except one spot that you would like to analyze. Finally, with keeping this state, you return to the bright image mode, and then you obtain a dark-field image. Here you find the part of the sample bright which corresponds to the diffraction spot you chose.

3. Electron Diffraction

The electron diffraction serves for determining crystallinity of a particular region of the sample, i.e., single crystalline or polycrystalline, crystal system, and orientation of the facet. For the measurement, first you select the region you want to analyze by using selected area aperture, and then you apply diffraction mode, whereby the current of intermediate lens would be zero and a diffraction pattern should appear on the screen. If the sample is amorphous, then a diffuse ring (halo pattern) would appear. Diffraction spots with two-, three-, four-, or sixfold symmetry would appear for single crystalline samples and coaxial circles (Debye-Scherrer rings) for polycrystalline samples. In rare cases such as some quickly chilled metal alloy samples of specific compositions, diffraction spots with five- or tenfold symmetry might be observed, suggesting a quasicrystal. For Debye-Scherrer rings, the following relationship holds between the distance R from the central point (corresponding to the position of transmitted electrons) to respective spots, the distance L between the sample and the detector (camera length), and the interplanar distance of the crystal d:

$$Rd = L\lambda \tag{16.4}$$

where λ is the wavelength of accelerated electrons. Furthermore, for samples belonging to the cubic crystal systems including face-centered cubic (fcc) and body-centered cubic (bcc), the following relationship holds between d and the lattice constant a:

$$d = \frac{a}{\sqrt{h^2 + k^2 + l^2}} \tag{16.5}$$

where (hkl) denote Miller indices. For single crystalline samples, the Miller indices are determined for linearly independent two diffraction spots. Then another vector being normal to these two vectors can be determined, which is called zone axis.

When the sample size is smaller than 10 nm, the electron diffraction would be very diffuse even for crystalline samples, so it might be difficult to determine its crystallinity. In such a case, high-resolution TEM images taken at $\geq 3 \times 10^5$ times of direct magnification would be helpful (Fig. 16.4). Here a lattice image is obtained due to interference between the transmitted electrons and diffracted ones. It allows you determining orientation of crystal facets and their Miller indices by measuring interplanar distances.

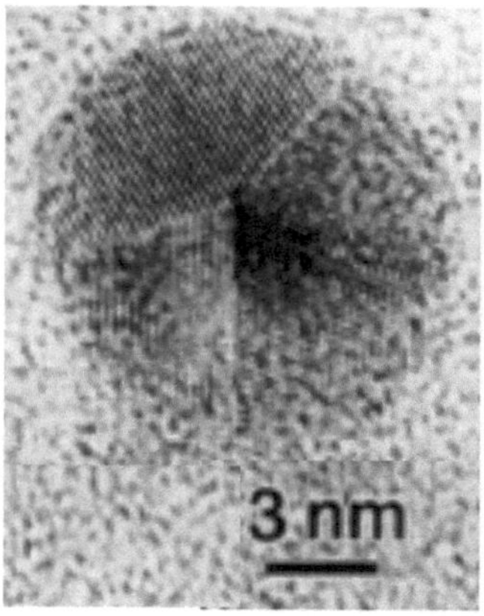

Fig. 16.4 High-resolution image of a multiply twinned Ag nanoparticle showing different orientations of lattices for each domain. Reprinted from Ref. [3]. Copyright 1993 American Chemical Society

16.3 Essentials and Tips

It is very important that the samples are well dried for both SEM and TEM observations. Besides, in SEM, the sample should be electrically conductive to obtain a high-contrast image. Therefore, for less- or non-conductive samples, heavy metals (Au, Pd, Pt) are vapor-deposited prior to the observation, if concomitant EDX analysis is not performed. When the sample is not sufficiently conductive, the image on the monitor irregularly trembles by charge-up effect, so high-quality images cannot be obtained. In TEM, the sample should be sufficiently thin to transmit electron beam; thus chemical or mechanical polishing should be performed for bulk materials, or sufficiently dilute samples are applied for powder samples.

On the other hand, we should be aware that the electron microscopy is a method of local analysis. Therefore, we always mind whether the obtained information is a representative one of the sample or a particular one for the observed area. It is also important to discuss the results in combination with other data obtained from global analysis, i.e., X-ray diffraction (XRD), X-ray photoelectron spectroscopy (XPS), and dynamic light scattering (DLS).

16.4 Useful Hints

For obtaining clear EM images, the sample concentration should not be too high, in particular for TEM observation.

When you want to analyze elemental composition of the sample, energy-dispersive X-ray analysis (EDX) or energy-dispersive X-ray spectroscopy (EDS) is a useful tool. The detector is often attached with SEM/TEM. The EDX has four modes, area analysis, line analysis, point analysis, and elemental mapping.

Very recently, new specimen holders have appeared which allow in situ EM observation [4], i.e., observing samples dispersed in liquid. It would be interesting to try such an observation technique.

Further Readings

1. C. Barry Carter, D. B. Williams (eds.), *Transmission Electron Microscopy – Diffraction, Imaging, and Spectroscopy* (Springer, Cham, 2016)
2. T. Pradeep, *NANO: The Essentials – Understanding Nanoscience and Technology* (McGraw Hill, New Delhi, 2008)

References

1. K. Torigoe, Y. Nakajima, K. Esumi, J. Phys. Chem. B **97**, 8304 (1993)
2. H. Liao, H. Zheng, Annu. Rev. Phys. Chem. **67**, 719–747 (2006)

Chapter 17
Gas Adsorption on Surface of Solid Materials

Takeshi Endo

Abstract Gas adsorption measurement provides essential information on surface properties of powder. In this chapter, theoretical background, process of measurement, and reading of obtained data are briefly introduced.

The adsorbate gas molecules are attracted by van der Waals force in physisorption. The relationship between adsorbed amount and activity (pressure), so-called adsorption isotherm, had been developed in early studies. Various isotherms have been proposed based on Langmuir monolayer adsorption theory, and the most important one is BET theory. Analysis of isotherm gives some important characteristics on solid surface, such as specific surface area, pore volume, pore size distribution, etc. The type of adsorption isotherm is classified into six types according to IUPAC criteria.

In actual measurement process, dry process in preparation of sample is very important to obtain accurate data. And measurement parameters should be set carefully taking various characteristics of samples into account. The obtained isotherm is analyzed by various analysis methods by computational calculations.

Keywords Gas adsorption and desorption · Adsorption isotherm · Pore size distribution · BET-specific surface area

17.1 Introduction

Interfaces between solid and gaseous phase are sometimes referred simply as "surface" and are very important to understand the behavior of liquid-solid interface. Basic characteristics of solid surfaces are usually characterized by interaction of gas molecules and solid surface. The specific surface area, one of the most important properties of solid surface, is estimated by measurement of adsorbed amount of gas molecules onto solid surface. In this chapter, outline of measurement technique of gas adsorption will be introduced.

T. Endo (✉)
Takeshi Endo, M Technique Co. Ltd., Izumi City, Japan

M. Abe (ed.), *Measurement Techniques and Practices of Colloid and Interface Phenomena*, https://doi.org/10.1007/978-981-13-5931-6_17

Atoms or functional groups on the most outer surface of solids show different behavior compared to inner part of the solids and attract surrounding molecules because of the asymmetric arrangement of surrounding atoms or groups. The phenomenon that molecules are attracted to the surface is called "adsorption." Attracted molecules are referred as "adsorbent," and the substance that attracts the surrounding molecules is referred to as "adsorbate."

When molecules adsorb onto the surface of solid materials, the potential energy decreases. And the system releases heat according to the decreased potential energy and stabilizes. This heat is referred as heat of adsorption and is always positive (exothermic).

Two attractive interactions are taken into account when the gaseous molecules adsorb onto solid surfaces. One is intermolecular-induced dipole-dipole interaction, so-called van der Waals force. Another is exchange of electrons between gaseous molecules and surface. The latter type of adsorption is chemisorption, and chemisorption always results in monolayer of adsorbed molecules. Chemisorption occurs in relatively high temperature and shows large amount of heat of adsorption, and the rate of adsorption is small due to its large activation energy. In contrary, adsorption induced by intermolecular attraction (van deer Waals forces) is referred as physisorption and observed in relatively low temperature. The heat of adsorption is very low compared to chemisorption, and the rate of physisorption is a large and reversible process. One example of physisorption is nitrogen molecules onto carbon; on the other hand, adsorption of oxygen molecules onto carbon surface is classified as chemisorption.

Many equations have been proposed to quantitative analysis of gas adsorption amount. The most simple one is Henry's adsorption isotherm, assuming that the adsorption amount simply be in proportional to the pressure of gas. The actual surface of solids is not a flat surface but complex geometry with unevenness and pores. An isotherm that is proposed by Langmuir can be applied to such complex surfaces. The isotherm was derived by consideration of monolayer adsorption dynamically.

At constant temperature T, the adsorbed volume of gas at standard condition V follows the equation shown below:

$$V = \frac{\mathrm{Kp}V_m}{1 + \mathrm{Kp}} \tag{17.1}$$

Here, K is a constant, p is an equilibrium pressure, and V_m is saturated adsorption amount, respectively. And Eq. (17.1) is converted to the equation below:

$$\frac{1}{V} = \frac{1}{\mathrm{Kp}V_m} + \frac{1}{V_m} \tag{17.2}$$

For example, adsorption amount of oxygen or carbon dioxide molecules onto silica gels follows this isotherm well.

The sites that gaseous molecules adsorb onto the surface distribute uniformly on the solid surface. One molecule occupies just one site in adsorption process, and one adsorption site can accept just one molecule for Langmuir's adsorption model. So once all adsorption sites were occupied by adsorbent, no further adsorption occurs by applying higher pressure. Such type of adsorption is referred as monolayer adsorption. The adsorption of Ar, N_2, CO_2, CO, and O_2 onto the activated carbon is categorized to monolayer adsorption.

BET adsorption isotherm was derived by Brunauer, Emmett, and Teller, and expanded Langmuir monolayer adsorption model to multilayer adsorption. At first at very low pressure, adsorbed molecular layer formed on the surface; after the monolayer was formed on the surface, the adsorbent gas adsorbs on the adsorbed gas molecules in monolayer and finally forms multilayers of adsorbed gas molecules on the surface. Such type of adsorption is called BET adsorption.

BET isotherm is expressed as equation below:

$$V = -\frac{V_m K_p P}{(P - P_0)\left\{\frac{(K_p - 1)P}{P_0} + 1\right\}} \tag{17.3}$$

Here, P_0 is saturated vapor pressure of adsorbent, p is relative pressure, and Kp is a constant. When $P_0 \gg P$ (at low pressure), the term p/p_0 is negligible compared to 1, and then Eq. (17.3) is equal to Langmuir's isotherm.

BET adsorption model is based on the conditions below in addition to the assumptions of Langmuir monolayer adsorption.

- The heat of adsorption is equal to heat of condensation of adsorbent molecules after second layer adsorption.
- When the p approaches to p_0, the adsorbed amount increases to infinity.
- No interactions between adsorbed molecules in each layer.

Adsorption on nonporous solid follows this type, and the main cause of attraction is van der Waals forces.

The adsorption process promoted by various interactions is described above. And adsorption isotherm was classified to some types, according to the properties of adsorbate such as pore size distribution, the morphology of pores, and chemical properties of the surface.

The classification of isotherms by IUPAC is shown in Fig. 17.1.

1. Type I: Monolayer Langmuir-type adsorption. Also applied to adsorbates with micropores (pores with diameter below 2 nm).
2. Type II: BET-type multilayer adsorption. Often seen in systems with strong interaction between surface and gas molecules.
3. Type III: Observed in systems with weak interaction between adsorbate and adsorbent and shows multilayer adsorption with inhomogeneous interactions.
4. Type IV: Observed with mesoporous surface with strong interaction between adsorbent and adsorbate. Often shows hysteresis loops with type H1–H4.

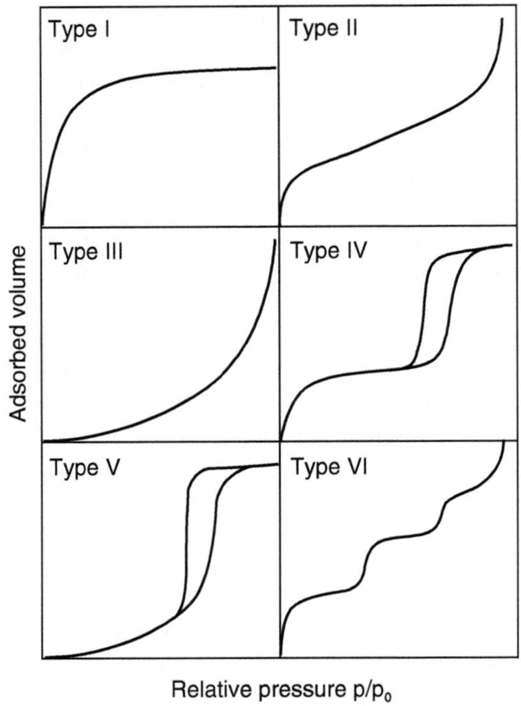

Fig. 17.1 The IUPAC classifications of adsorption isotherms

5. Type V: Observed with type III solid with mesopores. Also shows loops as type IV isotherm.
6. Type VI: Isotherm represents stepwise multilayer adsorption on a uniform homogeneous surface.

17.2 What You Get

The data directly obtained from measurement are the adsorbed amount in each relative pressure at a constant pressure. From an isotherm of nitrogen gas adsorption, we can obtain specific surface area (surface area per unit mass of adsorbent), pore size distribution, pore volumes, and so on. The morphology of pores can be estimated by comparison of hysteresis loop of the isotherm with IUPAC classification. For example, a solid with micropores shows steep rise at low p because of micropore condensation. The isotherm with hysteresis loop for H2 in IUPAC classification is sometimes connected with ink-bottle-shaped mesopores.

Distribution of metals on the surface can be estimated by measuring adsorption volumes with various gases which differently interact with a certain metal. Molecular probe method utilizes molecules with different sizes comparable to the micropores. The difference of adsorbed amount of different molecular cross sections gives information of smaller pores below limitation of normal pore size distribution calculation from nitrogen adsorption. Water vapor adsorption gives information about hydrophilicity of the surface. For nonporous spherical powders, the diameter of the particles can be obtained by BET.

17.3 Essentials and Tips

The measurement methods of adsorption volume of gas onto solid particles are classified as constant volume, constant flow, and weight measurement. The most popular one is the constant volume method. In constant volume method, the adsorbent gas is introduced to manifold with strictly defined volume at controlled temperature, and then the valve between the sample cell and manifold opened to introduce the gas into sample cell. The adsorption amount was calculated from the change of pressure using equation of the state of gas.

In actual measurement, ① the sample is set in the cell and weighed and thoroughly dried, ② cell is connected to the instrument and dried in vacuum if required, ③ input measurement conditions to PC of the instrument, ④ set liquid nitrogen to Dewar vessel, ⑤ start the measurement, and ⑥ analyze the isotherm.

17.4 Preparation of the Instrument

The measurement instrument (Fig. 17.2) is composed of a control panel (connected to PC), manifold, transducers (pressure meter), and vacuum pump for fine vacuum. Some connecting ports connect the instrument to adsorbent gases, roughing pump, and samples. Helium gas cylinder is connected to the instrument to flush the pipelines and measure the dead volume of sample cells. High purity of nitrogen gas is required to obtain high-quality data. Liquid nitrogen is used as coolant in nitrogen gas adsorption measurement. For the solids with very low specific surface area, krypton gas is often used.

17.5 Preparation of Sample

The typical diameter of sample cell is about 6–8 mm and 15–20 cm in length for powder samples. The cells are made of glass or quarts pipe with one end is closed. In typical instrument, the total surface area of introduced samples is recommended between 10 and 100 m^2. In case with too low surface area, it is hard to detect pressure change in adsorption, and with too large area, it takes very long time to reach equilibrium. When multiple samples were measured simultaneously, it is recommended that the pore volume or the total surface area is in close amount.

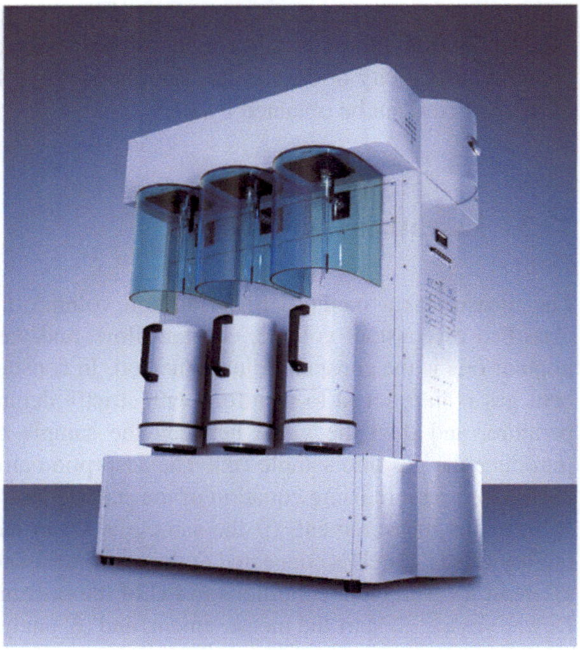

Fig. 17.2 An surface are and pore size analyzer (Photo credit Quantachrome a brand of Anton Paar)

The instrument measures only the pressure of gas. The weight of samples should be measured before or after the adsorption measurement with balance. Take care not to touch the sample cell with naked hand to avoid error in weighing.

Pretreatment of the sample (drying) should be done prior to the measurement. Residual moisture in the sample powder causes error in the pressure measurement because of the evaporation of moisture. The residual water also causes capping of pore. And evaporation of water sometimes causes the dispersion of powders resulting in the contamination of instrument or other samples. Samples are dried by heating in vacuum or gas flow. The operator should take care about the dispersion of powder during rapid decreasing of pressure and changes in crystalline and surface and morphological changes at high temperature. Dispersion of powder is caused by rapid expansion of water vapor, so slow pressure reduction or rough drying before introduction in vacuum is recommended. The amount of powder should not be excessive, and sample filter should be set in the open end of sample cell. For block samples, there is no need to care about dispersion, but these should be cut to introduce in the cell.

To avoid structural change by heating, the operator should set the pretreatment temperature below the phase transition temperature.

17.6 Setting of Parameter

Required properties are different for each instruments; the operator should input some parameters prior to the measurement. Select the adsorbate gas from the list in the PC, or if the gas is not in the list, input the values from literature. Set sample weight by the measured value by balance. Other detailed conditions, such as the duration of leak test, number of the point of measurement, criteria of judgment of equilibrium, and rate of gas introduction, should be set according to the instrument manual. In terms of the number of the point of measurement, 1 is enough for rough estimation of surface area, and for BET analysis, 7 points in 0.15–0.35 in p/p_0 are usually enough, and for pore size distribution analysis, some tens of measurement point are required in both adsorption and desorption process. For micropore analysis, many points in low p/p_0 should be measured. And sample drying and leak test should be conducted in strict condition.

It is recommended to determine these conditions after a rough measurement if possible.

17.7 What to Look Out for

Rough isotherm is obtained often when the total surface area is very low. In such case, one more measurement should be done with larger amount of sample, or use other adsorbate gas (e.g., Kr gas). Or each measurement does not reach equilibrium. In such cases, take longer duration to judge the equilibrium.

If the desorption blanch of isotherm is below adsorption blanch, leakage or evaporation of residual gas from powders may occur.

17.8 Understanding Your Data

A typical isotherm is shown in Fig. 17.3. The shape of the isotherm itself has meaning as described above; to obtain parameters such as specific surface area, etc., calculation with PC is required. For specific surface area, the BET method is most applied, and in labs, the adsorption instrument is merely called as "BET." Five or seven sets of adsorbed volume between $p/p0 = 0.15 - 0.35$ are input to BET equation. And monolayer adsorption amount is obtained by intercept and slope of BET plot. Then the cross section of molecules is multiplied with monolayer adsorption amount to give surface area. These calculation processes are done automatically by PC, but the fitting should be checked by the operator. The operator should select the line with highest correlation coefficient in $p/p0 = 0.15 - 0.35$ or around $p/p0 = 0$.

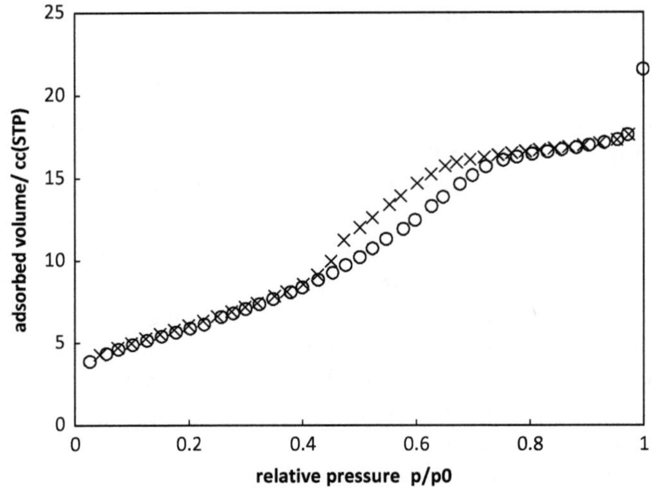

Fig. 17.3 Nitrogen gas adsorption-desorption isotherm of mesoporous titania

For pore size distribution, calculation method is different for mesopores (50–2 nm) and micropores (< 2 nm). Popular calculation methods are usually installed in PC, for example, BJH, DH, and CI method for mesopores and MP and t-plot method for micropores. In these days NLDFT (nonlocalized density function theory) calculation is installed in the analysis software, and this method is purely theoretical.

17.9 Useful Hints

Gas adsorption measurement is the most reliable method to obtain the value of specific surface area and most powerful method to investigate mesopores and micropores to which fluids other than gas cannot penetrate in because of the surface tension. TEM observation cannot be applied to evaluate inner pore of particles because of the incapable transmittance of electrons. For the analysis of pores, mercury or oil intrusion is suitable for larger sizes above 100 nm, absorption of positron beam is applied for smaller pores below 2 nm, and gas adsorption covers in between. By the combination of these analyses and electron microscopy, the researcher can obtain almost all morphological information of powders.

References

1. S. Kondo et al., *Chemistry of Adsorption* (in Japanese) (Maruzen, 2001)
2. M. Chikazawa et al., *Interfacial Chemistry* (in Japanese) (Maruzen, 2001)

3. B.C. Lippens, B.G. Linsen, J.H. De Boer, Studies on pore systems in catalysts I. The adsorption of nitrogen; apparatus and calculation. J. Catal. **3**, 32–37 (1964)
4. E.P. Barrett, L.G. Joyner, P.P. Halenda, The determination of pore volume and area distributions in porous substances. I. Computations from nitrogen isotherms. J. Am. Chem. Soc. **73**, 373–380 (1951)
5. K.S.W. Sing, D.H. Everett, R.A.W. Haul, L. Moscou, R.A. Pierotti, J. Rouquerol, T. Siemieniewska, Reporting physisorption data for gas/solid systems with special reference to the determination of surface area and porosity. Pure Appl. Chem. **57**, 603 (1985)

Chapter 18
Contact Angle Measurement for Solid Surface

Masahiko Abe

Abstract Contact angle measurement is used for the evaluation of the wettability of liquids to solid surfaces as a combination. Liquids with larger contact angles correspond to less wetting, while smaller means more wetting to solids. For example, the contact angle of water and clean glass surface is close to zero, but in contrast, it reaches up to about 110° for water- and paraffin-treated glass surface. This chapter deals with the measurement of contact angles to evaluate the wettability of water on solid surfaces.

Keywords Contact angle · Spreading wetting · Immersional wetting · Adhesional wetting · Critical surface Tension

18.1 Introduction

When water is dropped on a clean glass surface, it spreads rapidly as the glass gets wet, while when dropped on a paraffin-treated glass, the surface water makes droplets that stay at the point and the treated glass will not get wet (Fig. 18.1). The tangent line at P between the solid surface and the liquid surface gives an angle θ, defined as the contact angle. The contact angle for water and clean glass surface is close to zero but reaches about 110° for water- and paraffin-treated glass surface. As such, the contact angle represents the wettability of liquids and solids as a combination, where larger indicates less wetting and smaller indicates more wetting. There are two contact angles, the advancing contact angle θ_a and the receding contact angle θ_r as shown in Fig. 18.2. In general, θ_a is larger than θ_r, and this discrepancy is called hysteresis, which is caused by surface characteristics such as roughness, friction, and absorption.

M. Abe (✉)
Research Institute for Science and Technology, Tokyo University of Science, Noda, Japan
e-mail: abemasa@rs.noda.tus.ac.jp

M. Abe (ed.), *Measurement Techniques and Practices of Colloid and Interface Phenomena*, https://doi.org/10.1007/978-981-13-5931-6_18

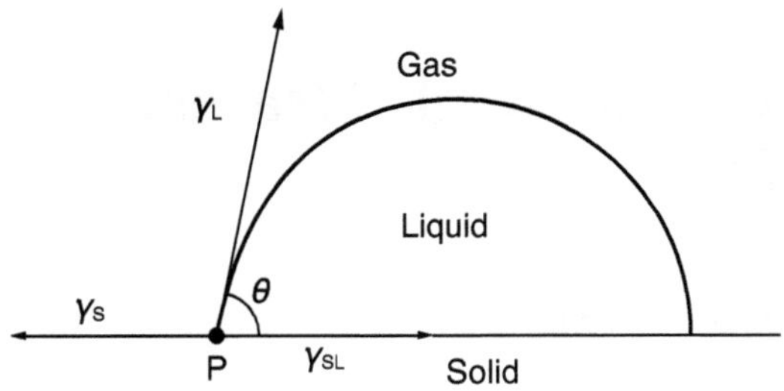

Fig. 18.1 Wettability of a liquid droplet on a flat solid surface. The contact angle (θ) is determined by the balance in horizontal direction of three surface and interfacial tensions of γL, γS, and γSL

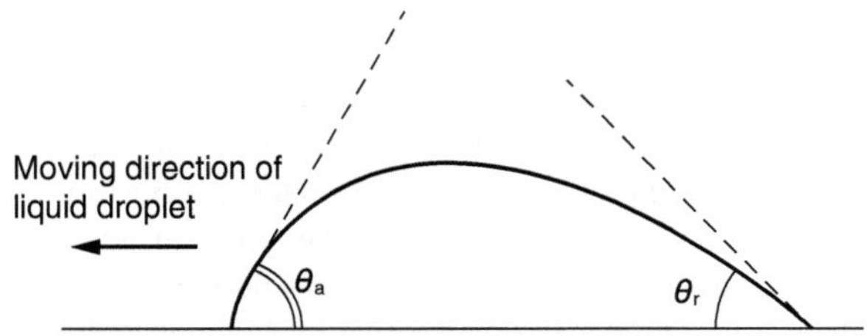

Fig. 18.2 Advancing contact angle and receding contact angle

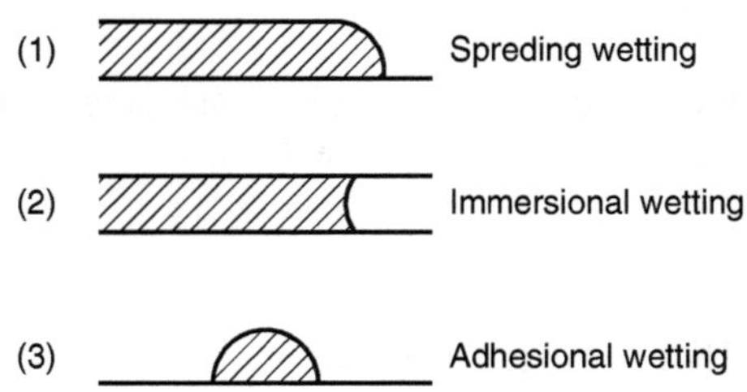

Fig. 18.3 Wetting types

18.2 Wetting and Immersion

When a small amount of liquid is dropped on a surface, it will either spread or stay as a droplet with a certain contact angle. There is an interface between the solid and air which is generally called the surface of solid. In order for the liquid to contact the surface, the liquid has to push air away from the surface to make a new interface between the liquid and solid. Wetting is a phenomenon that creates this liquid/solid interface. There are three types of wetting, namely, spreading wetting, immersional wetting, and adhesional wetting as shown in Fig. 18.3.

18.2.1 Spreading Wetting

Spreading wetting is a type of wetting where water or alcohol spreads over a glass surface, and its work of spreading (W_s) is expressed by Eq. (18.1):

$$W_s = \gamma_s - \gamma_l - \gamma_{sl} \tag{18.1}$$

where γ_s, γ_l, and γ_{sl} are the surface tension of solid and liquid and interfacial tension between the surface and liquid, respectively. Larger W_s corresponds to higher wettability.

18.2.2 Immersional Wetting

Immersional wetting is a type of wetting where water penetrates into filter paper, and its work of spreading is expressed by Eq. (18.2):

$$\mathrm{Wi} = \gamma_s - \gamma_{sl} \tag{18.2}$$

18.2.3 Adhesional Wetting

Adhesional wetting is a type of wetting where mercury is dropped on a glass surface, and its work of spreading is expressed by Eq. (18.3):

$$W_s = \gamma_s + \gamma_l - \gamma_{sl}\theta \tag{18.3}$$

where the contact angle is determined by the balance in the horizontal direction of the three surface and interfacial tensions, γ_s, γ_l, and γ_{sl}, as shown in Fig. 18.1, which is expressed by Young's equation (Eq. 18.4) as follows:

$$\gamma_s = \gamma_{sl} + \gamma_l \cos\theta \tag{18.4}$$

By applying Young's equation (Eq. 18.4), each work of spreading can be expressed as follows:

$$\mathrm{Ws} = \gamma_l(\cos\theta - 1) \tag{18.5}$$

$$\mathrm{W}_i = \gamma_l \cos\theta \tag{18.6}$$

$$W_a = \gamma_l(\cos\theta + 1) \tag{18.7}$$

The wetting can be determined as:

When:

$\theta = 0°$, $W_s \geqq 0$ spreading wetting occurs.
$\theta \leqq 90°$, $\mathrm{Wi} \geqq 0$ immersional wetting occurs.
$\theta \leqq 180°$, $\mathrm{W}_a = 0$ adhesional wetting occurs.

In summary, solid surfaces can be categorized as hydrophilic surfaces and hydrophobic surfaces. To make a surface non-wettable to both water and oil, the solid surface tension (γ_s) must be smaller than both γ_{water} and γ_{oil}.

18.3 Measurement of the Contact Angle

The most important key for accurately measuring contact angles is to prepare a clean surface by extensive cleansing. There are two typical ways for the measurement.

18.3.1 θ/2 Method

This method is applicable when a liquid is dropped on surfaces and spontaneously shapes as part of a sphere, depending on its surface tension (γ_l) as shown in Fig. 18.4. This picture can be taken using a CCD camera, and the contact angle θ can be determined by the following equation (Eq. 18.8):

$$\tan\theta_1 = \frac{h}{r} \rightarrow \theta = \arctan\frac{h}{r} \tag{18.8}$$

where r is the diameter of the droplet and h is the height of drop. If the drop is small enough in the range of a few μl, the effect of gravity can be ignored, and the droplet shape can be treated as part of a sphere.

18.3.2 Tangent Line Method

As shown in Fig. 18.5, points L_1, L_2, and L_3 on the drop arc are each treated as part of a circle. The center of the circle (M) can be determined, and then the tangent line

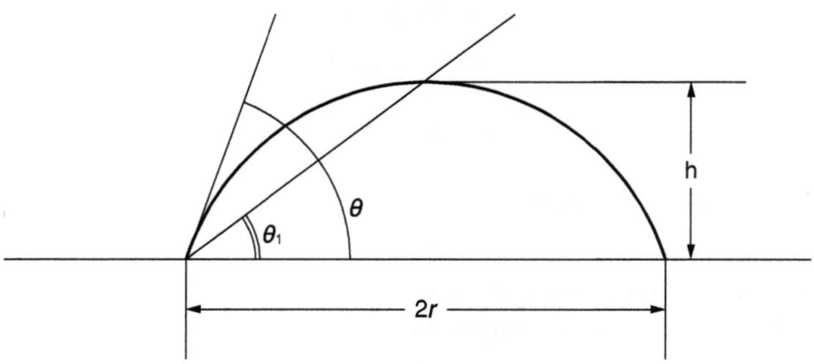

Fig. 18.4 Illustration of θ/2 method

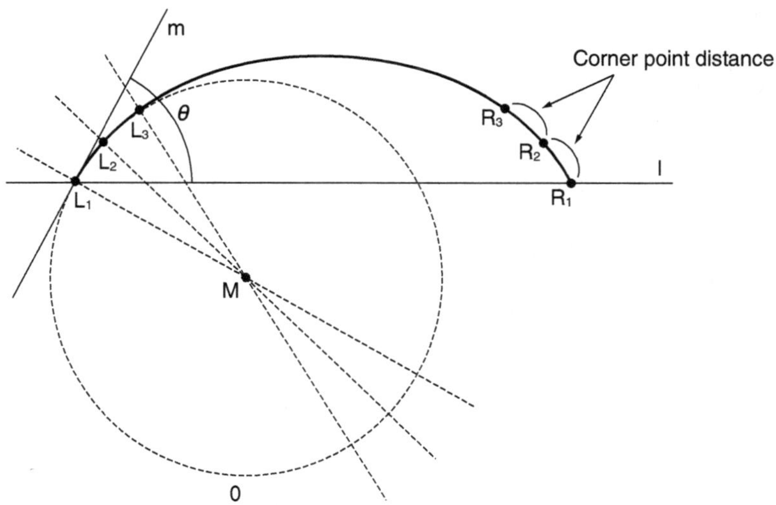

Fig. 18.5 Illustration of tangent line method

(M) at L_1 can be drawn as shown in Fig. 18.5. The contact angle θ at the left side of the droplet can be obtained between the tangent line m and the surface line l. The right-side contact angle can be obtained as well in the same manner from the points R_1, R_2, and R_3.

The contact angle θ by the θ/2method is the average of both sides of the droplet as θ is calculated by width and height of the droplet. As such, the tangent line method would be a better choice when the shape of the droplet is asymmetric such as when on an uneven surface or when the droplet is ascending or receding, as shown in Fig. 18.2.

18.4 Determination of Critical Surface Tension

As mentioned above, wetting phenomena can be determined by the intrinsic surface tension of liquids (γ_l) and solids (γ_s). Although γ_l can be relatively easily measured, experimental determination of γ_s is rather difficult. Utilization of critical surface tension of solids (γ_C) is a method to estimate γ_s from the contact angle of various liquids on the target solid surface. The following are details of the method to determine γ_C.

Table 18.1 shows the surface tension of various liquids [1]. Here, Type A liquids involve liquids whose surface tension (γ_l) derives only from London dispersion force (γd) such as n-alkane (saturated hydrocarbon). Type B liquids include liquids with London dispersion force (γd) and dipol-dipol interaction force (γb) such as n-alkene (un-saturated hydrocarbon). Type C liquids are liquids with (γd), (γb) and hydrogen bonds (γh) such as 1-alnanol (fatty alcohol). Determination of the critical surface tension (γ_C) of solids is summarized as (1) dropping a liquid which γ_l is known, as

Table 18.1 Surface tension of standard liquids at 30 °C

Type	Standard liquids		Surface tension (mN/m)
Type A liquid*1	Dodecane	$C_{12}H_{26}$	25.0
	Tridecane	$C_{13}H_{28}$	25.7
	Tetradecane	$C_{14}H_{30}$	26.5
	Pentadecane	$C_{15}H_{32}$	26.9
	Hexadecane	$C_{16}H_{34}$	27.4
	Heptadecane	$C_{18}H_{36}$	27.9
Type B liquid*2	1-Dodecene	$C_{12}H_{24}$	27.7
	1-Tetradecene	$C_{14}H_{28}$	28.3
	1-Hexadecene	$C_{16}H_{32}$	29.4
	1-Octadecene	$C_{18}H_{36}$	30.1
Type C liquid*3	1-Hexanol	$C_6H_{13}OH$	25.7
	1-Heptanol	$C_7H_{15}OH$	26.6
	1-Octanol	$C_8H_{17}OH$	27.2
	1-Nonanol	$C_9H_{19}OH$	28.0
	1-Decanol	$C_{10}H_{21}OH$	28.4
	1-Dodecanol	$C_{12}H_{25}OH$	29.1

*1: which contains component of London dispersion force (γ^{d})
*2: which contains both γ^{d} and component of polar force (γ^{P})
*3: which contains γ^{d}, γ^{P}, and component of hydrogen bonding force (γ^{d})

Fig. 18.6 Relationship between cos*θ* and the surface tension of the liquids for PSt-C6F13 (Rf ratio = 16%) film at 30 °C

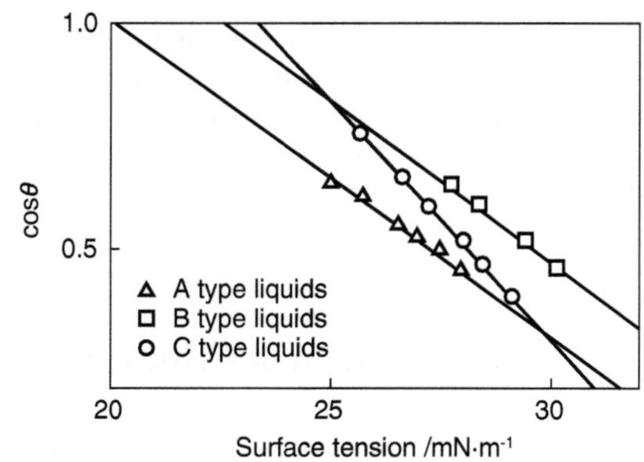

shown in Table. 18.1, and measuring the contact angle to get cosθ and (2), as shown in Fig. 18.6 [2–3], plotting each cosθ value for the serial liquids (Types A, B, and C) against the surface tension as Zisman plot. The critical surface tension for each type of liquid (γ^{A}_{C}, γ^{B}_{C}, and γ^{h}_{C}) is determined at the intersection between each line and $\cos\theta = 1$ as shown in Fig. 18.6. Each contribution of molecular forces for γ_C is calculated as follows [4–6]: (1) The London dispersion force γ^{d}_{C} corresponds to γ^{A}_{C}. (2) Polar parts (γ^{P}_{C}) can be obtained by subtracting γ^{B}_{C} from γ^{A}_{C} ($\gamma^{P}_{C} = \gamma^{A}_{C} - \gamma^{B}_{C}$). (3) Hydrogen bond contribution can be determined by subtracting γ^{B}_{C} from γ^{C}_{C} ($\gamma^{h}_{C} = \gamma^{C}_{C} - \gamma^{B}_{C}$). The sum of these elements (γ^{t}_{C}) gives the contact angle of

Table 18.2 Components of critical surface tension of PSt-$CF(CF_3)O\{CF_2(CF_3)0\}_mC_3F_7$ (mm = 0,1, 2) films at 30 °C

Sample	Rf ratio	γ_c^d (mN/m)	γ_c^P (mN/m)	γ_c^h (mN/m)	γ_c^t (mN/m)
PSt	0%	30.3	4.4	1.0	35.7
m = 0	2%	22.0	4.1	1.1	27.2
	3%	21.6	3.9	1.1	26.6
	5%	19.9	3.6	1.1	24.6
	10%	18.4	3.1	1.2	22.7
m = 1	4%	21.0	2.9	1.5	25.4
	6%	19.2	3.2	1.6	24.0
	13%	12.4	2.9	0.9	16.2
	43%	12.3	2.9	1.0	16.2
m = 2	3%	14.0	3.5	1.1	18.6
	9%	10.1	3.0	1.1	14.2
PSt-C_6F_{13}[a]	16%	20.5	2.3	0.7	23.5

[a]Data from lit [6]

specific solid surfaces as shown in Table 18.2 as $\gamma^t_C = \gamma^d_C + \gamma^P_C + \gamma^h_C$. Table 18.2 shows the data for these values of a perfluoroalkanoyl-polystyrene surface [7].

The water and oil repellant property of perfluoroalkanoyl-polystyrene surface is shown in Table 18.2. The degree of fluoro substitution to hydrogen at the alkane moiety remarkably decreases γ^t_C, which reaches 14.2 nm/m and is smaller than Teflon (18 nm/m), and this is why these solids show excellent repellency both for water and oil.

General features of wetting can be found in the recent review article by Tsujii [8].

References

1. K. Ogino, M. Abe, M. Morikawa, H. Sawada, M. Nakayama, J. Oil Chem. Soc. **40**, 1115 (1991). Japanese
2. W.A. Zisman, J. Phys. Chem. **58**, 260 (1954)
3. W.A. Zisman, *Contact angle, wettability and adhesion*, Adv. Chem. Ser., 43 (American Chemical Society, Washington, DC, 1964)
4. T. Hata, Polymer **17**, 594 (1968). Japanese
5. M. Kitasaki, T. Hata, J. Adhes. Soc. Jpn. **8**, 131 (1972). Japanese
6. M. Abe, M. Morikawa, K. Ogino, H. Sawada, T. Matsumoto, H. Shimada, N. Matsubayashi, A. Nishijima, Japanese. J. Jpn Soc. Colour Mater. **65**, 475 (1992)
7. M. Morikawa, M. Abe, K. Ogino, H. Sawada, T. Matsumoto, H. Shimada, N. Matsubayashi, A. Nishijima, Japanese. J. Jpn Soc. Colour Mater. **65**, 612 (1992)
8. K. Tsujii, Wetting and surface characterization: chapter 23, in *Cosmetic Science and Technology: Theoretical Principles and Applications*, ed. by K. Sakamoto et al., (Elsevier, Oxford, 2018), pp. 373–388

Chapter 19
Quality and Treatment of Water for Experiment Use

Masahiko Abe

Abstract In most cases, water for physiochemical experiments is used after percolating and boiling to remove impurities such as particles, organics, colloids, microorganisms, etc. To remove bacteria from ultrapure water is arguably the most difficult process. For example, gram-negative bacteria such as *Pseudomonas fluorescens* and *Pseudomonas putida* are found in purified water with a TOC (total organic carbon) of 60 ppb, while *Pseudomonas stutzeri* is found in further purified water with a TOC of 5 ppb.

This chapter deals with types of water appropriate for using in experiments.

Keywords Ultrapure water · Gram-negative bacteria · Pyrogen · TOC · Ion-exchange resin treatment

19.1 Impurity of Water for Experiment Use

Manufacturing processes use a large amount of ultrapure water every day [1]. Although called "ultrapure," this water contains trace amounts of impurities such as particles, organics, colloids, microorganisms, etc. Impurities in water are classified as shown in Table 19.1. When the impurities are chemical material, they can be removed by physical or chemical treatment to levels of a few ppt, but it is difficult to completely remove microbiological metabolites. The problem for chemists dealing with water purification is that it is impossible to reach a purity of 100%. The reason is that microorganisms can survive even under extremely high or low pH or high temperature. Even once bacteria have been temporally killed, water can be recontaminated.

When water is used as a solvent for an experiment, especially for an extra accurate measurement, its reproducibility is rather low. To measure nm-sized solubilized

M. Abe (✉)
Research Institute for Science and Technology, Tokyo University of Science, Noda, Japan
e-mail: abemasa@rs.noda.tus.ac.jp

M. Abe (ed.), *Measurement Techniques and Practices of Colloid and Interface Phenomena*, https://doi.org/10.1007/978-981-13-5931-6_19

Table 19.1 Classification of impurities in water

Impurities	Electrolyte	Strong electrolyte	Dissolved substance	Na^+, K^+, Ca^{2+}, Mg^{2+}, Cl^- NO_3^-, SO_4^{2-} etc
			Suspended matter	Metal hydroxide oxide sulfide carbonate etc
		Weak electrolyte	Dissolved substance	Carbonic acid Organic acid Organic base etc
			Suspended matter	Heavy metal salt of arganic acid Insoluble silicate etc
	Non electrolyte	Dissolved substance	Organic compound Dissolved gas etc	
		Suspended matter	Insoluble organic compound Microorganism Plankton Alga Oily matter etc	

materials in water, contamination of bacteria gives false-negative and irreproducible results, so extra care should be taken.

19.2 Importance of the Quality of Water Used in Experiments

Here, we focus on gram-negative bacteria such as *E. coli* [2]. Pyrogenous materials mainly exist at the outer layer of gram-negative bacteria cell membranes and cause strong fevers for mammals and other animals. More endotoxin (or pyrogen) was found in summer than during the dry season in tap water taken from Tone River, as shown in Fig. 19.1. The compositions of the elemental atoms in the bacteria as dry matter are C 50%, Na 6.4%, Ca 2.9%, Mg 1.5%, and K 1.2%. Bacteria (counted as 1×10^4/ml) can survive under a TOC (total organic carbon) over 1 μg/l, so this TOC is a critical parameter for water quality.

The number of bacteria in the tap water treated by distillation or ionic resin is shown in Table 19.2. Regardless of these treatments, there were no bacteria found immediately after the treatment though significant numbers were counted after only 2 days of storage, and this is assumed to be from bacterial growth utilizing tiny amounts of organic and other related materials as nutrients [3].

The growth profiles of bacteria collected from the water are shown in Figs. 19.2 (for the twice distilled water and water for injection) and 19.3 (water treated by mixed bed of cation-exchange and anion-exchange resin). The bacteria count for twice distilled water reached 10^4 after 48 h, while for the water treated with ion-

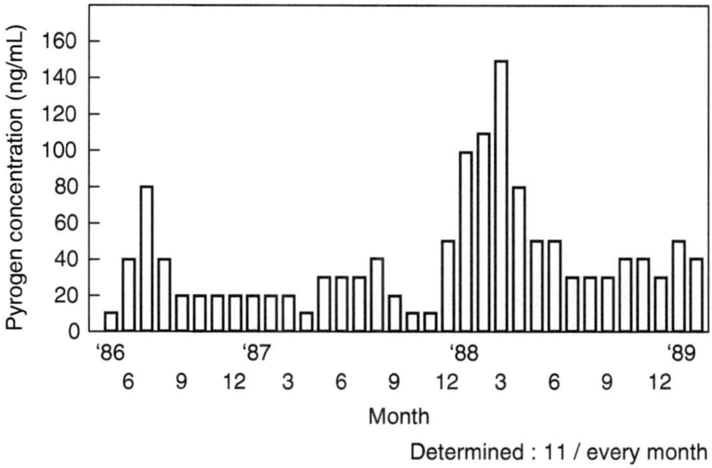

Fig. 19.1 Monthly change of pyrogen level in tap water

Table 19.2 Change in bacteria number in treated water

	Once distilled water		Twice distilled water		Deionized water	
Treated water	Fresh	Stored	Fresh	Stored	Fresh	Stored
Bacterial number (cell/100 mL)	0	15	0	50	0	25

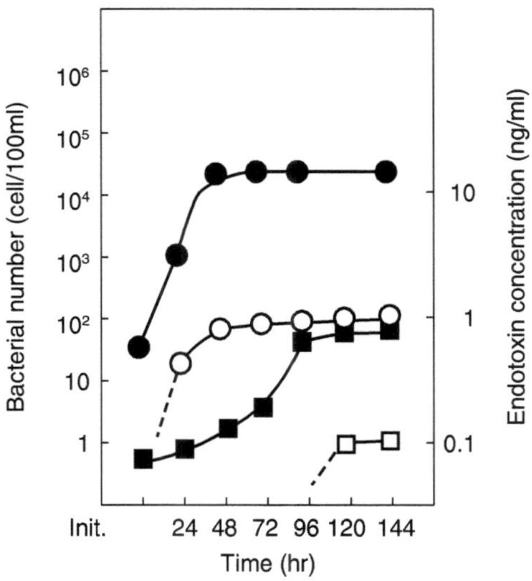

Pure water	Conductivity (µs/cm)	TOC (ppb)	Bacterial number	Endotoxin concen
Twice distilled water	1.7	218	●	○
Water for injection	0.9	138	■	□

Fig. 19.2 Changes of bacterial number and endotoxin concentration in pure water at 25 °C

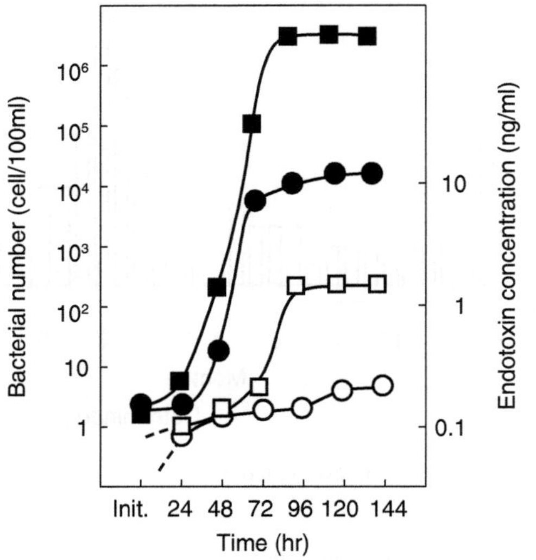

Deionized water	Conductivity (μs/cm)	TOC (ppb)	Bacterial number	Endotoxin concen
Amb.-200C/ IRA-900	1.0	245	●	○
Amb.-200C/ IRA-900	1.3	169	■	□

Fig. 19.3 Changes of bacterial number and endotoxin concentration in deionized water at 25 °C

Table 19.3 Average particle size of reference endotoxin[a] in water for injection

Endotoxin	Average particle size (nm)
LPS from *E. coli* 055: 85	157 (141–166)
LPS from *E. coli* UKT-B[b]	212 (200–219)
LPS from *E. coli* UKT-B[c]	171 (161–178)
	(500 ng/ml)

[a]LPS = endotoxin
[b]Include bovine serum albumin
[c]Include mannitol

exchange resin bacterial growth abruptly increased after 72 h, and the number of bacteria reached 10^6 for the water with TOC 245 ppb. However, reducing TOC prevented bacterial growth as shown in Fig. 19.3. This result indicates that there was serious bacterial growth for ion-exchanged water (TOC 245 ppb) in the wash bottle after 4 days of storage, so special attention should be paid for the quality of stored water.

The particle size of endotoxin regarded as a sphere shape in the water for injection (TOC c.a. 150–170 ppb) is around 160 nm to 220 nm, quite large compared to water molecules as shown in Table 19.3. The ζ potential of bacteria in water is negative, around -30 mv as shown in Table 19.4, so contamination of bacteria and its residue interferes with physical and chemical experiments since the water became a heterogeneous solvent. The effect of water purification for bacteria was investigated as

Table 19.4 Zeta potential of various kinds of bacteria in water

Kinds of bacteria		Zeta potential (mV)
A. ruhrandii	IAM 12600	-35 ~ -45
A. faecalis	IAM 12586	-35 ~ -40
A. putrefaciens	IAM 12079	-20 ~ -30
E. coli	IAM 1268	-40 ~ -50
P. aeruginosa	JCM 2776	-30 ~ -35
P. diminuta	JCM 2778	-30 ~ -40
S. marcescens	IFO 12648	-10 ~ -20

Table 19.5 Identification of bacteria from pure water

	TOC 60 μgC L		TOC 5 μgC L
Characteristics	Bacterial A	Bacterial B	Bacterial C
Gram stain			
Catalase test	+	+	+
Oxidase test + + +			
Nitrate reduction test	–	–	+
H2S generation test	–	–	+
Acid formation from sugars			
D-Glucose	+	+	+
Fructose	+	+	+
Maltose			
Xylose	+	+	+
Mannitol	–	–	–
Lactose	–	–	–
Saccharose	+	–	–
Trehalose	+	–	Not tested
Growth on MacConkey agar	+	+	+
Growth at 42 °C	–	–	+
Pigment formation	+	+	+
Identification	*Pseudomonas fluorescens*	*Pseudomonas putida*	*Pseudomonas stutzeri*

shown in Table 19.5. *Pseudomonas fluorescens* and *Pseudomonas putida* were found in purified water with a TOC of 60 ppb, while *Pseudomonas stutzeri* was found in the further purified water with a TOC of 5 ppb [3].

19.3 Types of Water Appropriate for Experiments

General classification of water is shown in Table 19.6. Industrial water is prepared by first separating large floating materials from river water, and then dirt and sand are separated by sedimentation. Tap water is prepared successively by separating

Table 19.6 Wide-ranging variety of water

	River water	Water for industrial use	Water supply	Pure water	Ultrapure water
Turbidity	<15	10 ~ 15	<2	<1	<1
Resistivity	0.001 ~ 0.01 MΩ·cm	0.001 ~ 0.01 MΩ·cm	0.2 ~ 0.02 MΩ·cm	0.2 ~ 15 MΩ·cm	16 MΩ·cm
Fine particle	Uncountable	Uncountable	Several 10^3 ~ Several 10^4/cm^3	Several 10^2/cm^3	<10^2/cm^3
Bacteria	Uncountable	Uncountable	Several 10^3/cm^3	Several 10^2/cm^3	<10^2/cm^3
Organics	>15 ppm	1 ~ 15 ppm	1 ~ 5 ppm	1 ppm	<0.5 ppm

Resistivity of ideal pure water 18.25 MΩ • cm (25 °C)

floating materials, and coagulant treatment is applied for sedimentation and is sterilized. Pure water is prepared from tap water or well water by ion-exchange resin treatment or distillation to remove electrolytes and other contaminants. Ultrapure water is prepared by further treatment to achieve its specific conductivity to as close as 18.2 MΩcm, the theoretical value for pure water [3, 4]. This ultrapure water is indispensable for various cutting-edge technologies such as energy, electronics, etc. The requirement for the quality of ultrapure water is becoming more severe due to the increasing improvement of product purity, functionality, ultrafine processing, ultra-sterilization, and pollution control.

In order to fulfill such requirements, intensive combination of unit technologies such as ion exchange, membrane filtration, and UV radiation is developed for the production of ultrapure water. Typical production processes consist of pre-purification by sedimentation and filtering colloidal turbid materials, followed by reverse osmotic membrane treatment to remove salts, organic molecules, fine particles, and microorganisms solubilized or dispersed in the pretreated water. Solubilized gases such as CO_2 and oxygen are removed by a degas tower. Residual ions are then treated by ion-exchange resin to make pure water. This water is sterilized and treated by ultrafiltration and reverse osmotic filtration, processes called final filtration, to remove tiny debris of ion-exchange resin, microorganisms, and ultrafine particles to make ultrapure water. Wastewater from this final purification process is recycled back to the pretreatment point.

Here, we will explain sterilization by ion exchange [5]. Ion-exchange resins are categorized as anion- and cation-exchange resins, and it is known that H^+-type cation-exchange resin does not have sterilizing effects. Chronological count of bacteria for four types of OH^- exchange resin treatments is shown in Fig. 19.4. For all four resins, decrease in bacteria was observed when adding resin regardless of the number of their reactivation. No recurrence of bacteria was observed for IRA-458 and IRA-958 after the count reached 0/ml. For IRA-402BL and IRA-900, reduction of bacteria count is gradual, whereas IRA-458 and IRA-958, especially IRA-402BL, did not reach to 0/ml during the test period. As such, OH^-

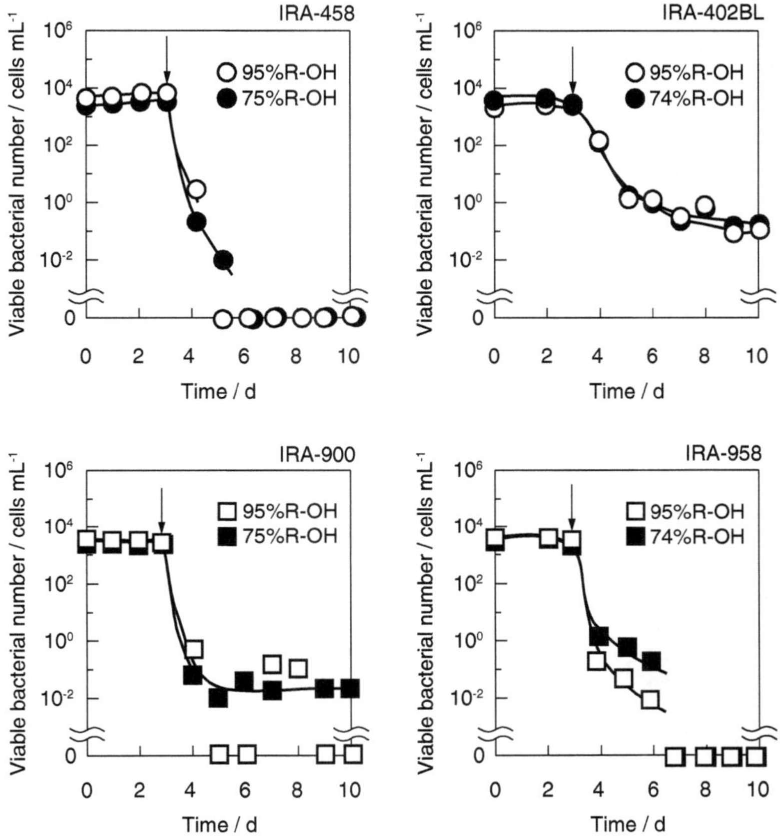

Fig. 19.4 Disinfection capabilities of acrylic SBAERs and styrenic SBAERs in OH- form with different regenerated levels. IRA-458, IRA-958, IRA402BL, and IRA-900 were added on the third day

type has sterilizing effect, and acrylate type (IRA-458 and -958) is stronger than styrene type (IRA-900 and -402BL). These resins were custom-made for this study.

Figure 19.5 shows the effect of the combination of anion- and cation-exchange resins against single anion-exchange resin. As clearly seen in Fig. 19.5, OH^- anion-exchange resin combined with H^+-type cation-exchange resin showed a superior sterilizing effect compared to OH^- type used alone, and especially the best effect was found at 3/1 ratio [5]. The reason for the advantages of the combination use is shown in Fig. 19.6. For OH^- anion-exchange resin only, the bacteria adsorb to resin by electrostatic attraction and hydrophilic interaction and then contact the OH^- as counterion to kill bacteria. Sterilized bacteria remains on the ion-exchange resin. The reason why OH^- ion sterilizes bacteria is unclear, though abrupt pH change might be a cause of killing bacteria. The number of bacteria in the vicinity of OH^- anion-exchange resin surface is larger than the bulk, and coagulation of bacteria might be enhanced to result in the acceleration of its absorption to the hydrophilic resin

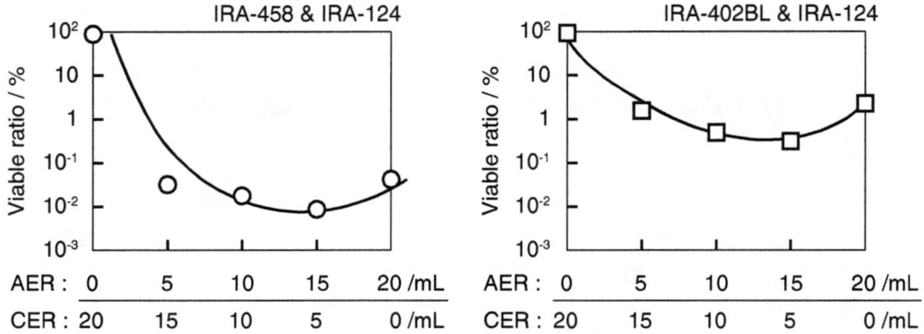

Fig. 19.5 Dependence of the volume ratio of the IERs in OH- form to the IERs in H- form on bactericidal capabilities, after incubation for 24 h at 35 °C: (**a**) IRA-458 and IR-124; (**b**) IRA-402BL andIR-124

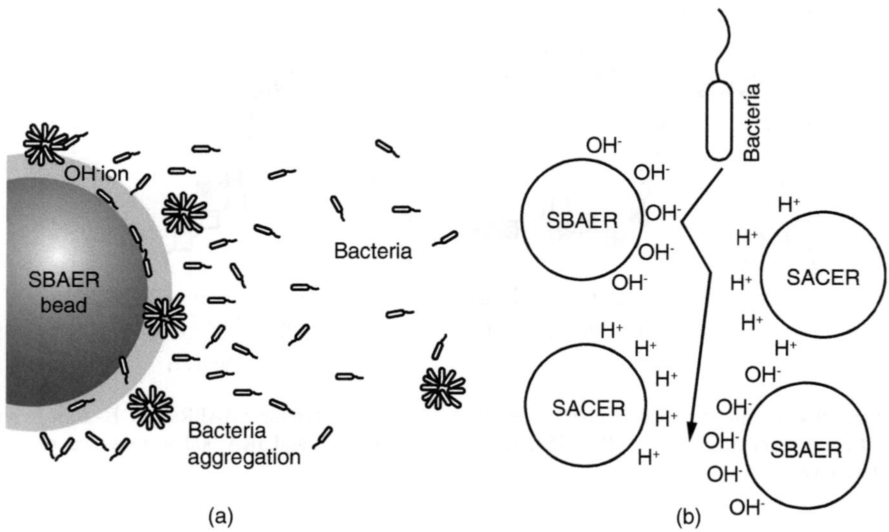

Fig. 19.6 Disinfection mechanism of bacteria with (**a**) OH- ions and (**b**) both OH- ions on the surface of the ion-exchange resins

surface. Thus, more hydrophilic acrylate resin enhanced the sterilization effect compared to less hydrophilic styrene resin. For the mixed bed with OH^- anion-exchange resin and H^+ cation-exchange resin, bacteria contact both OH^- and H^+ at the resin surface alternatively, so that sudden pH change is assumed to have caused the effective sterilization.

Standardization of water for analytical laboratory use can be found as ISO3696 [6].

References

1. Nanae Matsuda, *Master's Degree Theses* (Tokyo University of Science, 1996)
2. W. Agui, Y. Kurachi, M. Abe, K. Ogino, J. Antibact. Antifung. Agents **16**, 313 (1988)
3. N. Matsuda, W. Agui, T. Togou, H. Sakai, K. Ogino, M. Abe, Colloids Surf. B Biointerfaces **5**, 279 (1966)
4. A. Iverson, J. Phys. Chem. **68**, 515 (1964)
5. N. Matsuda, W. Agui, K. Ogino, N. Kawashima, H. Sakai, M. Abe, Colloids Surf. B: Biointerfaces **7**, 91 (1966)
6. Water for analytical laboratory use- specification and test methods, ISO3696, First edition 1987-04-15

MIX
Papier aus verantwortungsvollen Quellen
Paper from responsible sources
FSC® C105338

If you have any concerns about our products,
you can contact us on
ProductSafety@springernature.com

In case Publisher is established outside the EU,
the EU authorized representative is:
Springer Nature Customer Service Center GmbH
Europaplatz 3, 69115 Heidelberg, Germany

Printed by Libri Plureos GmbH
in Hamburg, Germany